6th International Conference on Green Energy and Sustainable Development (GESD 2023)

E3S Web of Conferences Volume 478 (2023)

Shenyang, China
18-19 November 2023

Editors:

Zhihui Tan
Lin Liu

ISBN: 978-1-7138-8819-2

Printed from e-media with permission by:

Curran Associates, Inc.
57 Morehouse Lane
Red Hook, NY 12571

Some format issues inherent in the e-media version may also appear in this print version.

For additional information, please contact EDP Sciences – Web of Conferences
at the address below.

EDP Sciences – Web of Conferences
17, Avenue du Hoggar
Parc d'Activité de Courtabœuf
BP 112
F-91944 Les Ulis Cedex A
France

Phone: +33 (0) 1 69 18 75 75
Fax: +33 (0) 1 69 28 84 91

contact-edps@webofconferences.org

Additional copies of this publication are available from:

Curran Associates, Inc.
57 Morehouse Lane
Red Hook, NY 12571 USA
Phone: 845-758-0400
Fax: 845-758-2633
Email: curran@proceedings.com
Web: www.proceedings.com

TABLE OF CONTENTS

https://doi.org/10.1051/e3sconf/202447800001

Preface: 6th International Conference on Green Energy and Sustainable Development (GESD 2023)

This proceeding includes original and peer-reviewed research papers from the 6th International Conference on Green Energy and Sustainable Development (GESD 2023), which held on November 18-19, 2023 in Shenyang City, China.

GESD 2023 is organized by Research Center of Engineering and Science, co-organized by Chongqing University, Asian Union of Information Technology, sponsored by CCCC Second Harbour Engineering Company Ltd., Chishun Chemical.

The aim of GESD 2023 was to provide a platform for researchers, engineers, academicians, as well as industrial professionals, to present their research results and development activities in Green Energy and Sustainable Development. It provides opportunities for the delegates to exchange new ideas and application experiences, to establish business or research relations and to find global partners for future collaboration. This two-day conference was being organized in invited keynote talks and three Parallel Sessions - Green Energy & Sustainable Energy and Power; Resource Extraction, Utilization & Sustainable Development; Environmental Sustainability and Green Technologies.

All full paper submissions to this GESD would be sent to at least two reviewers and evaluated based on originality, technical or research content, correctness, relevance to conference, contributions, and readability. The full paper submissions were chosen based on technical merit, interest, applicability, and how well they fit a coherent and balanced technical program. GESD 2023 received 86 manuscripts. And 30 submissions had been accepted by our reviewers. The Committees of GESD 2023 invested great efforts in reviewing the papers submitted to the conference and organizing the sessions to enable the participants to gain maximum benefit.

Zhihui Tan
Lin Liu

Conference Organizing Committee

Statement of Peer review

In submitting conference proceedings to *Web of Conferences*, the editors of the proceedings certify to the Publisher that

1. They adhere to its **Policy on Publishing Integrity** in order to safeguard good scientific practice in publishing.
2. All articles have been subjected to peer review administered by the proceedings editors.
3. Reviews have been conducted by expert referees, who have been requested to provide unbiased and constructive comments aimed, whenever possible, at improving the work.
4. Proceedings editors have taken all reasonable steps to ensure the quality of the materials they publish and their decision to accept or reject a paper for publication has been based only on the merits of the work and the relevance to the journal.

Title, date and place of the conference

GESD 2023
November 18-19, 2023
Shenyang City, China

Proceedings editor(s):

Zhihui Tan
Lin Liu

Date and editor's signature January 9, 2023

Study on Cracking Causes of One 500 kV SSY Type Equipment Clamp

Jianjun Zheng[1,*], Yong Wei [2], Xiaokai Liu [1], Kuo Han[1]

[1] Inner Mongolia Power (Group) Co., Ltd., Inner Mongolia Power Research Institute Branch, Hohhot 010020, China
[2] Inner Mongolia Helin Power Generation CO.LTD, Hohhot 010020, China

Abstract. One SSY type equipment clamp serviced in a 500 kV station was found to be cracked. In this paper, the crcaked cause and mechanism is investigated by means of macroscopic morphology observation, fracture surface morphology observation, X-ray digital detection, metallographic observation. The results indicated that the main reason for the fracture of the equipment clamp this time is that the terminal board is a cast part, which contains a large number of casting defects such as porosity and shrinkage, resulting in a decrease in its effective cross-sectional area and load-bearing capacity. In addition, insufficient modification treatment of the terminal board led to insufficient toughness due to needle or rod-shaped eutectic silicon in the organization. Finally, under the combined action of dynamic load and self weight caused by the swinging of the wire in the wind, the terminal board was brittle fractured along the bolt hole with obvious stress concentration effect.

Keywords: Al; clamp; fracture; casting; defect

1. Introdution

Due to the rapid development of the power grid, high-voltage transmission lines are becoming more and more concentrated, and the use of equipment wire clamps is also increasing[1-2]. The quality of this component is always related to the safety and stability of the power grid operation. The equipment wire clamp normally consists of a fastening twisted wire part and a connecting part. The former is a gland and wire groove structure or tubular structure, while the latter is a flat terminal board structure connected to the electrical equipment through bolts[3-4]. The typical morphology of SSY type equipment clamp is shown in Figure 1. The wiring tube is generally made of industrial pure aluminum hot-rolled, and the terminal board is made of cast aluminum alloy, wihch are connected by welding. The wire clamp only bears the gravity of itself and a portion of the wire without excessive stresses.

In actual production, due to factors such as processing costs, production schedule requirements, and uneven personnel skill levels, casting defects are easily formed during the terminal board castings of aluminum alloy equipment clamps, and the welds between the terminal board and the wiring tube are prone to welding defects such as incomplete fusion and incomplete penetration[5-8]. These defects not only increase the energy loss during the power transmission process, but also greatly reduce the strength and load-bearing capacity of the clamps, causing cracks or even breaking druing the operation of equipment wire clamps[9-10]. Therefore, conducting failure analysis of equipment wire clamps is of great significance for the safe and stable operation of the power grid.

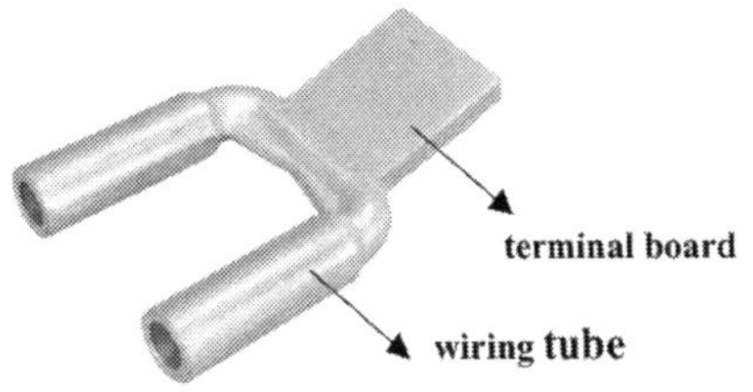

Figure.1 SSY type equipment clamp

During the power outage maintenance process of a 500 kV substation, the terminal board of CVT side SSY type equipment wire clamp was found to be broken, and the and the model of the fractured wire clamp is JS/Z-WT-2021-1002. In this paper, the causes and mechanisms of fractured clamp is studyed by a series of tests and analyses, and proposed corresponding governance measures are also supplied.

* Corresponding author: dkyzjj@163.com

2. Experimental results and analysis

2.1 Experimental method

Firstly, external macromorphology observation of the equipment clamp is carried out,to determine whether it has original defects, mechanical damage, arc burns, etc. Subsequently, according to the requirements of DL/T 1785-2017[11], the overall perspective testing of the equipment clamp is carried out by using ERESCO 65MF4 type X-ray digital imaging testing system (DR), to determine whether its manufacturing quality meets the standard requirements. Then the microstructure of the equipment clamp terminal board is detected by using the Axio Observer.Alm type metallographic microscope, to determine whether its metallographic structure is normal according to the requirements of GB/T 13298-2015[12].The micro morphology of the cross section of the equipment clamp terminal board is analyzed by using the S-3700N type scanning electron microscope (SEM) comprehensive test system, to determine its fracture mechanism according to the requirements of JY/T 010-1996[13]. The hardness of the equipment clamp wiring terminal sample is conducted by the THBC-3000DA Vickers hardness tester. The diameter of the cemented carbide ball is 5 mm, the test load is 2500 N.

2.2 Macroscopic morphology observation and analysis

It can be seen from Figure.2(a) that the fracture is located at the bolt hole of the wire clamp terminal board. Due to the existence of the bolt hole, the bearing area of this area is significantly reduced compared to other areas, which is prone to stress concentration cracking. The fracture surface is rough without obvious deformation, showing a silver white metallic luster and typical brittle fracture characteristics, as shown in Figure.2(b). In addition, there are also a large number of small casting defects such as porosity and shrinkage in the cross-section.

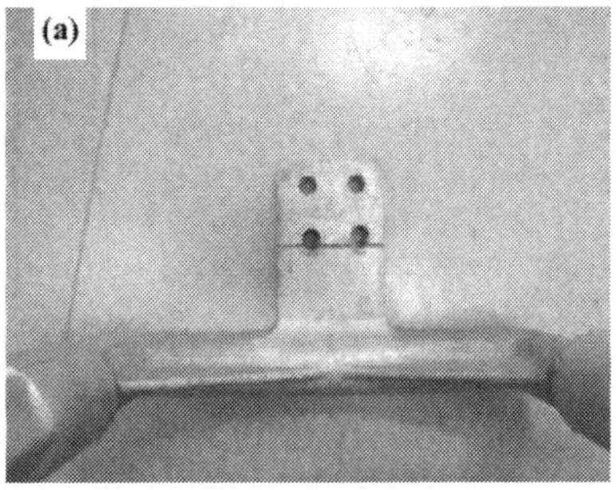

Figure.2 Macromorphologies of the broken equipment wire clamp integral (b) fracture

2.3 Observation and analysis of fracture surface

After ultrasonic cleaning, the fracture microstructure of the equipment clamp is analyzed using scanning electron microscopy (SEM), and the results are shown in Figure 3. It can be seen that the fracture surface is flat and presents a wood grain like feature. There are a large number of dendritic nodular shrinkage casting defects in the cross-section, which is consistent with the typical brittle fracture characteristics of cast aluminum alloys.

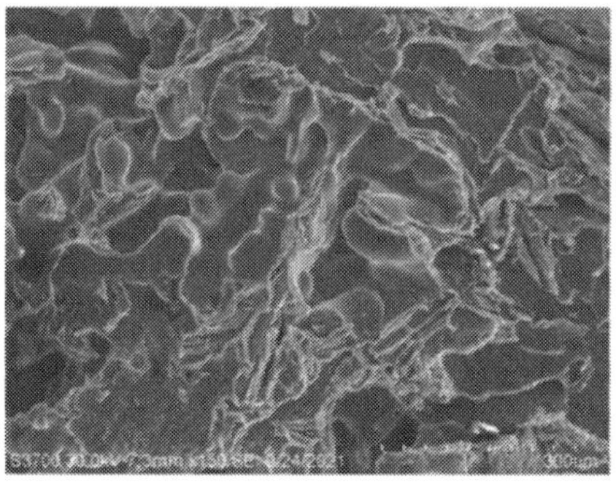

Figure.3 Fracture morphology of the equipment clamp wiring terminal

2.4 X-ray digital detection and analysis

Then DR detection on the broken equipment clamp is adopt, results as shown in Figure 4. It can be seen that there are a large number of casting defects such as porosity and shrinkage in the terminal board of the equipment clamp.

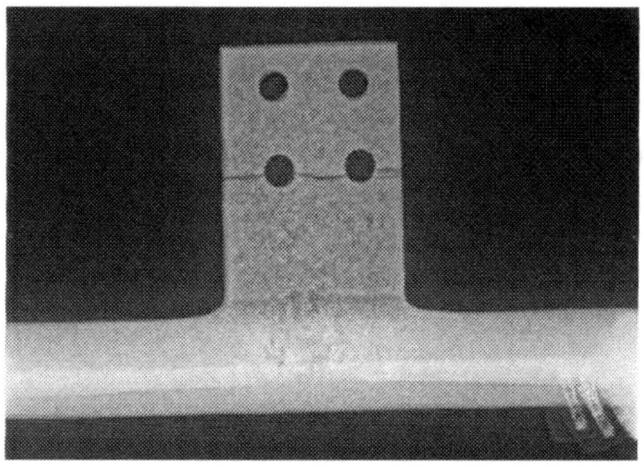

Figure.4 Perspective morphology of the broken equipment clamp

2.5 Microstructure detection and analysis

Figure.5 reveals the microstructures of the broken equipment clamp terminal board. It can be found that the terminal board matrix is mainly consisted of α(Al) phase+needle or rod shaped eutectic silicons, accompaning by a large number of casting porosity or shrinkage defects, which is principally due to the insufficient metamorphic treatment. Related studies have shown that needle or rod-shaped eutectic silicon can lead to insufficient toughness of the terminal board[14-16].

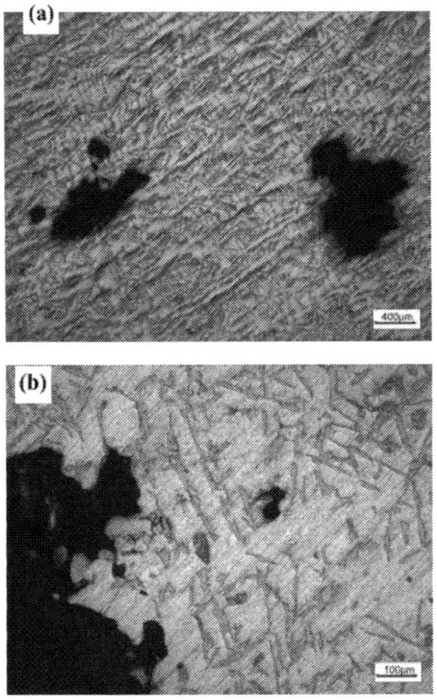

Figure.5 Microstructures of the fractured equipment clamp wiring terminal

2.6 Hardness detection and analysis

The hardness testing results show that the average measured hardness of the equipment clamp wiring terminal sample is 50.1 HBW, indicating that the measured hardness of the clamp meets the standard requirements according to GB/T1173-1995[17].

3. Analysis of fracture mechanism

According to the macroscopic morphology analysis results, the equipment clamp is cracked form the bolt hole of the terminal board. Due to the existence of the bolt hole, the bearing area of this area is significantly reduced compared to other areas, which is prone to stress concentration cracking. The fracture surface is rough and shows a silver white metallic luster, with typical brittle fracture characteristics. In addition, there are also a large number of small casting defects such as porosity and shrinkage in the cross-section. From the analysis result of the fracture morphology of the equipment clamp wiring terminal, it can be seen that the fracture of the terminal board of the equipment clamp is flat and presents a wood grain like feature. Besides, there are a large number of dendritic nodules like shrinkage casting defects in the cross-section, which is consistent with the typical brittle fracture characteristics of cast aluminum alloys. From the metallographic analysis results, the metallographic structure of the broken equipment wire clamp terminal board is consisted of α phase+needle or rod-shaped eutectic silicons, accompanied by a large number of casting porosity or shrinkage defects, which is a product due to insufficient modification treatment. Needle or rod-shaped eutectic silicon can lead to insufficient toughness of the terminal board and is highly prone to brittle fracture. At last, it can be seen that there are a large number of casting defects such as porosity and shrinkage in the broken equipment clamp terminal board from the DR perspective image, resulting in a decrease in the effective cross-sectional area and load-bearing capacity of the terminal board.

4. Conclusions and suggestions

Based on the aboved analysis results, the main reason for the fracture of the terminal board of SSY type equipment clamp serviced in one 500 kV transformer substation is that the terminal board of the equipment clamp is a cast part, which contains a large number of casting defects such as porosity and shrinkage, resulting in a decrease in its effective cross-sectional area and bearing capacity. At the same time, insufficient modification treatment of the terminal board led to insufficient toughness due to needle or rod-shaped eutectic silicon in the organization. Finally, under the combined action of dynamic load and self weight caused by the swinging of the wire in the wind, the terminal board brittle fractures along the bolt hole with obvious stress concentration effect. At last, according to the requirements of 6.2.5 listed in 《Technical supervision code for metal in power grid equipment》 (Q/GDW11717-2017)[18], the terminal boards of equipment clamps and tension clamps should not use casting technology. It is recommended that the terminal boards of newly replaced equipment clamps should be manufactured using forging technology. Secondly, the inspection of equipment clamps serviced in the substation should be strengthen, promptly recording and handling any abnormal situations such as cracking of the terminal board to avoid the recurrence of similar types of fracture failure.

Acknowledgments

This research was financially supported by 2022 Youth Science and Technology Project of Inner Mongolia Power Research Institute Branch (Grant No. 2022-QK-09).

References

1. XU Xian, WU Guozhong, BAO Yanrong, et al. Analysis on fracture fault of aluminum device clamp in substation[J]. Casting Forging Welding, 2012, 41(7): 191-195.

2. DONG Jin, ZHONG Xiangyuan, AN Yinghong. Fracture of copper and aluminum middle-sized device clamp in substation[J]. Electric Power, 2010, 43(12):27-30.
3. HONG Yicheng, CAI Jianbin. Analysis on fracture reasons of rectangular hanging plates of hanging fittings in the 220 kV electric transmission line[J]. PTCA (PART A :PHYSTEST), 2017, 53(5): 361-364.
4. Cause of cracking on power grid terminal connector[J]. PTCA (PART A :PHYSTEST), 2021, 51(7): 66-73.
5. XIA Xiaojian, ZHANG Konglin, LIN Deyuan. Failure analysis on cracking of an aluminum alloy flangefor current transformer[J]. PTCA (PART A :PHYSTEST), 2017, 53(10): 754-760.
6. DONG Zhongli, HU Ping, LUO Xue. Fracture failure analysis on cast aluminum alloy splint of highvoltage isolation switch static contact[J]. Mechanical and Electrical Engineering Technology, 2017, 46(2): 76-78.
7. WANG Ruoming, ZHAN Maji, JI Kun, et al. Failure reason of tension clamp for high voltage electricity transmission[J]. Materials for Mechanical Engineering, 2017, 41(3): 112-116.
8. XU Xuexia, OUYANG Jie, FENG Yanting, et al. Analysis on crack failure of terminal connector[J]. Casting Forging Welding, 2011, 40(13): 160-163.
9. XU Jie. Reasons and improvement measures for 110 kV equipment wire clamps[J]. Yunnan Electrical Power, 2006, 34(3): 45.
10. LUO Hong jian, GONG Zhuo, WANG Jionggeng, et al. Failure analysis of aluminum terminal connector[J]. Journal of Yunnan Normal University Journal, 2018, 38(5): 44-48.
11. DL/T 1785-2017. Directive for X-ray digitalradiography of electric power equipment[S]. National Energy Administration, 2018.
12. GB/T 13298-2015. Inspection methods of microstructure for metals[S]. General Administration of Quality Supervision, Inspection and Quarantine of the People's Republic of China, 2016.
13. JY/T 010-1996. General rules for analytical scanning electron microscopy[S]. State Education Commission, 1997.
14. DING Wenbin, FANG Wen, ZHANG Chaolei, et al. Failure analysis of double conductor aluminum equipment wire clamp for transmission tower[J]. Special Casting Nonferrous Alloys, 2023, 43(3): 359-362.
15. LUO Changsen, LUO Hong, ZENG Xianguang. Fracture analysis of Al alloy ZL102 polishing machine ring[J]. Special Casting Nonferrous Alloys, 2013, 33(3): 285-288.
16. HAN Jiceng, HE Jian, LIN Deyuan. Fracture reason of L-type junction board of capacitor voltage transformer accessory in a substation[J]. Materials for Mechanical Engineering, 2020, 44(7): 98-102.
17. GB/T1173-1995. Casting aluminium alloys[S]. The State Bureau of Quality and Technical Supervision, 1996.
18. Q/GDW11717-2017. Technical supervision code for metal in power grid equipment[S]. State Grid Corporation of China, 2018.

Evaluation of integrated geological engineering reservoir reconstruction in Block X of Bozi

Fengyuan Yang [2], Yang Xiao [1,3], Ju Liu [2], Shiyong Qin [2], Dengfeng Ren [2], Fen Peng [2], Zhonghui Ma [1]

[1] Chengdu University of Technology, Chengdu, Sichuan 610059, China
[2] PetroChina Tarim Oilfield Company, Korla, Xinjiang 841000, China.
[3] Chengdu Technology Sun Energy Technology Co. Ltd., Chengdu, Sichuan 610059, China

Abstract: The geological structure of the reservoir in Bozi-Dabei area is complex, and the longitudinal difference of rock mechanical properties of the reservoir is large. Most of the current evaluation techniques after reservoir reconstruction are based on the fitting of construction pressure based on the logging curve of a single well, so as to analyze the fracture morphology and the geometric dimensions of length, width and height after pressure. It is difficult to explain the fracture morphology and geometric parameters correctly because the interpretation methods used do not comprehensively consider the influence of natural fractures, 3D geological properties and 3D geomechanical properties. In order to explain the artificial fracture morphology more accurately, the target layer ant body is generated by using 3D original seismic data, and the natural fracture model of three different reservoir types in Bozi X block is generated, which provides geological basis for the integrated demonstration. By combining dynamic and static data and multidisciplinary methods, a one-dimensional rock mechanics profile model is established, and a three-dimensional geomechanical model of three different reservoir types in Block X of Bozi is established by combining geological model and deep diagenesis theory. The matching relationship between horizontal stress, fault strike and natural tensile fracture distribution is comprehensively analyzed and evaluated, providing a geomechanical basis for the evaluation of integrated reservoir reconstruction. Integrated geological model and geomechanical model are closely combined with engineering parameters to establish an integrated true three-dimensional artificial fracture model. On this basis, the optimization research and design of the parameters of displacement, liquid volume and sand ratio are completed. The optimal flow rate of Class I reservoir 6-7 (m^3·min-1), Class II reservoir 5-6 (m^3·min-1), Class III reservoir 4-5 (m^3·min-1), Class I reservoir 600-700 m^3, Class II reservoir 600-700 m^3, Class III reservoir 500-600 m^3, Sand ratio of 15%-20% for Class I reservoir, 15%-20% for class II reservoir and 20%-25% for class III reservoir are the optimal construction parameters, which provide guidance for field construction.

Keywords: Bozi Block; Tight reservoir; Geological engineering integration; Ant tracking; Fracturing parameter optimization

1. Introduction

The reservoirs in the Bozi-Dabei area are the second and third lithologic sections of the Cretaceous Bashijiqike Formation, and this set of reservoirs is widely distributed regionally, with the second lithologic section being a subphase of the braided delta front and the third lithologic section being a subphase of the fan delta front. The Lower Cretaceous of Bozi-Dabei area is a fractured dense sandstone reservoir formed under the depositional system of fan delta and braided river delta, and the area is dominated by feldspar lithic sandstone, followed by lithic feldspar sandstone, and the particle-size is dominated by medium and fine grains; the reservoir space is dominated by intergranular pores (primary intergranular pores, intergranular dissolution expanded pores) and dissolved pores in grains, with a very small number of micropores and fissure pores; the reservoir fractures is well developed, and the main ones are upright and high-angle shear fractures. The geological structure of the area is complex, the mechanical properties of reservoir rocks vary greatly in the longitudinal direction, and the local stress disturbance characteristics are obvious because of the existence of a large number of upright and high-angle shear fractures in the plane. Most of the current techniques for post reservoir reforming assessment are based on fracturing construction curves for G function, horner function and other characteristic curve analysis, or construction pressure fitting based on single-well logging curves to analyze the post-pressure fracture morphology and geometric dimensions of length, width, and height. And the interpretation methods used basically do not consider the effects of natural fractures, 3D geological attributes and 3D geomechanical attributes, and do not consider the single fracture model of fractured dense sandstone reservoir modification of multi-fracture multi-stage stress perturbation characteristics, which makes it difficult to correctly interpret the morphology of post-pressure fracture and geometric parameters.

In recent years, the existing design of fracturing parameters is mainly based on numerical simulation of single wells and analytical methods. Gou et al used numerical simulation to optimize the fracturing parameters based on a fine geological model of the reservoir. Qu et al established the gas reservoir model and fracture model, and optimized the ground fracturing parameters. Zhong used an analytical method to optimize the fracturing parameters based on basic seepage theories such as reset potential and potential superposition principle. Therefore, this paper establishes a block geomechanical model based on the regional stress changes and deep diagenesis characteristics, while considering the dynamic ground stress field changes coupled with faults, natural fractures and dynamic changes in production. Based on this, an integrated artificial fracture model containing geological model and geomechanical model is constructed for the evaluation of post-pressure artificial fracture morphology and geometry, and the longitudinal and planar expansion law of artificial fracture in typical well areas has been studied to comprehensively complete the optimization study of fracturing for reservoir modification and provide theoretical basis for the preference of regional fracturing construction process and optimal design of parameters.

2. Modeling of natural fractures in the target layer

Using the 3D original seismic data body to generate the target layer ant body. First of all, through the original seismic data body construction smoothing processing, to reduce the noise interference, and enhance the continuity of the seismic reflection wave in the same phase axis. Then the ANOVA technique is used to obtain the variance body for the seismic data body that has been processed by tectonic smoothing to highlight the discontinuous information of the seismic data. Finally, the ant tracking parameters are tested several times to determine the optimal parameter values suitable for the study area, and the discontinuous traces in the ant tracking technology that satisfy the preset fracture conditions in the other differential body are tracked to generate the natural fracture model of the ant body of the three different reservoir types in Block X of Bozi, as shown in Fig. 1.

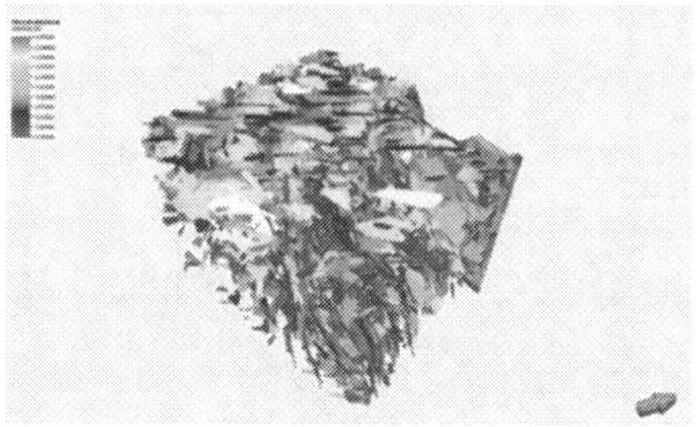

(a) Type I reservoir

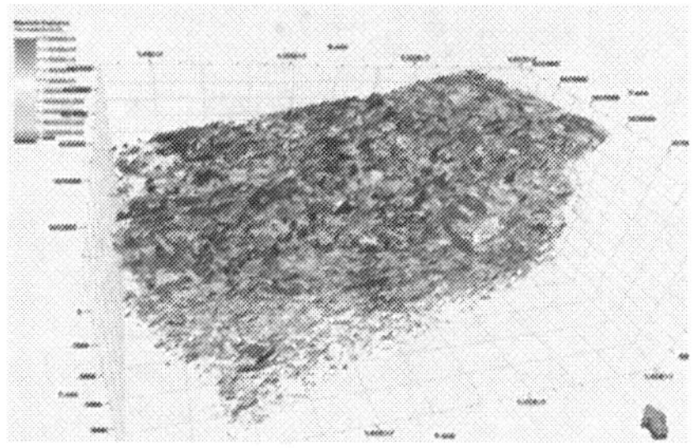

(b) Type II reservoir

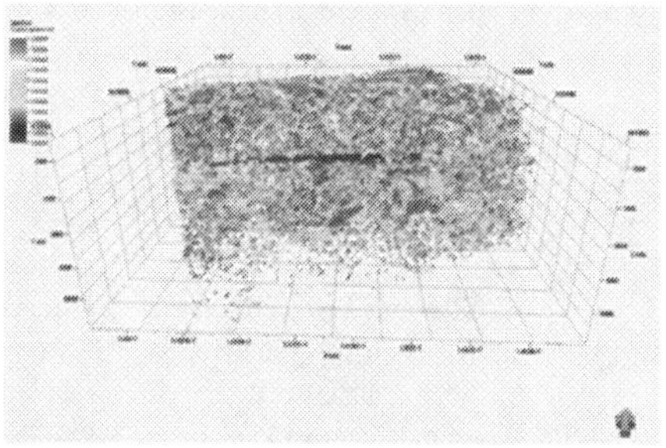

(c) Type III reservoir

Fig. 1 Natural fracture model of ant body in 3 different reservoir types in Block X of Bozi (a, b, c)

3. Mechanical environment construction and stress-strain matching calibration of deep-seated stratum

The initialized 3D geomechanical attribute body is created by filling in the overlying strata, basement rocks and surrounding rocks on the basis of the geological model. Horizontal stress is applied around the area to simulate the process of deep-seated stratum formation, fit the results of mechanical interpretation of well points, correct the basic parameters of geomechanics, and preliminarily obtain a static three-dimensional geomechanical model under the present-day ground stress conditions. Load the stress environment, faults and natural fracture network system, simulate the deep rock formation process, respectively establish the artificial fracture-containing deformable saturated reservoir flow-solid coupling seepage model, artificial fracture-containing deformable saturated reservoir flow-solid coupling elastic deformation of rock model, artificial fracture-containing deformable saturated reservoir

coupling temperature field model, synthesize the initial 3D geostress model, the results of the numerical simulation of reservoirs derived from the 3D pore pressure and temperature field, and comprehensively establish the dynamic geostress model. temperature field derived from the numerical simulation results of the reservoir, and establish the dynamic geostress model.
Based on fault strike of the block, the horizontal stress direction is analyzed, and the initial stress direction of the model is preliminarily generated by combining with the regional stress study. Among them, positive faults and tensional fractures have a certain matching relationship with the direction of the maximum horizontal principal stress based on their historical causes, especially the direction of the tensional natural fractures formed within each period is basically consistent with the direction of the maximum horizontal stress of the current period. On this basis, the dynamic geomechanical model is formed by fitting and adjusting the rock mechanical parameters through multiple stress evolutions, and finally reaching the maximum unbalanced force convergence.

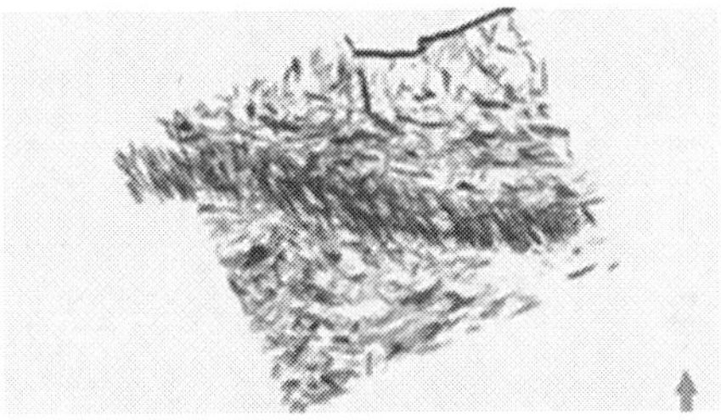

(a) Type I reservoir

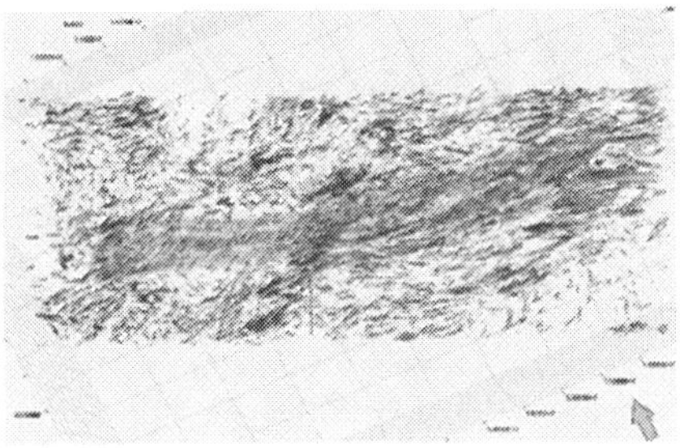

(b) Type II reservoir

(c) Type III reservoir

Fig. 2 Faults, natural fractures and maximum horizontal direction matches in 3 different types of reservoirs in Block X of Bozi

4. Evaluation and optimization study of artificial fractures after fracturing

The expansion law of artificial fracture is not only related to geological conditions, but also linked to the construction parameters in the field. In order to optimize the best construction parameters with a view to achieving the best fracturing effect, based on the established natural fracture model and 3D geomechanical model, a true 3D artificial fracture model of Pre-Tertiary stratum in well A is established, and the optimization study of its fracturing construction parameters is carried out by using Single Factor Method (as shown in Table 1). It is necessary to study the artificial fracture expansion law of different reservoir types in Bozi block and finally get the optimized fracturing design parameters.

Table 1 Simulation range of artificial fractures parameters

engineering parameters	**range**		
	Type I reservoir	**Type II reservoir**	**Type III reservoir**
displacement of fracturing fluid/ m^3/min	3、4、5、6、7、8	2、3、4、5、6、7	2、3、4、5、6、7
volume of fracturing fluid/ m^3	300、400、500、600、700、800		
sand proportion/%	5、10、15、20、25、30		
Proportion of prepad fluid/%	20、30、40、50、60、70		
perforation thickness/m	2、5、8、11、14、17、20		

4.1 Displacement of fracturing fluid

Based on the established fracturing model to simulate the effect of displacement, the displacement was set to 2, 3, 4, 5, 6, 7 and 8 (m^3/min), respectively, to analyze and study the changes of fracture length, height, and width, as well as the changes of the support area and the transformation volume under different displacements. With the increase of displacement, the overall trend of fracture length, height, support area and transformation volume shows a rising trend, and the transformation effect of the artificial fracture network is reflected in the size of the values of the fracture length and support area, when the displacement reaches 6-7 (m^3/min) (Type I reservoir), 5-6 (m^3/min) (Type II reservoir), 4-5 (m^3/min) (Type III reservoir), the artificial fracture length, support area and transformation volume size tend to be stable, and the comprehensive site economic factors are considered, so the optimal displacement is considered to be 6-7 (m^3/min) for type I reservoir, 5-6 (m^3/min) for type II reservoir, and 4-5 (m^3/min) for type III reservoir. The artificial fracture morphology at optimal displacement for the 3 types of reservoirs are shown in Fig. 3.

(a) Type I reservoir

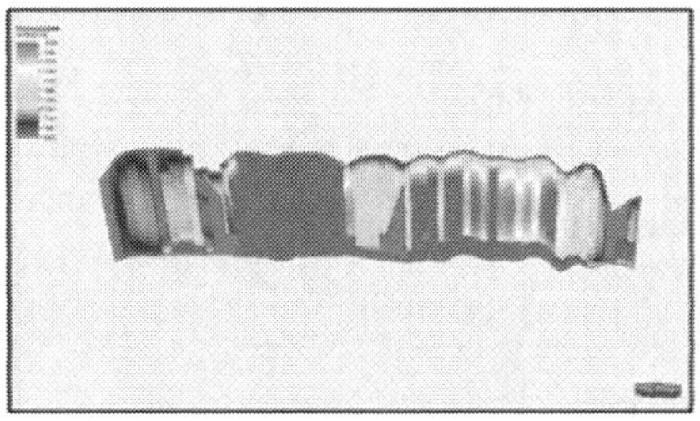

(b) Type II reservoir

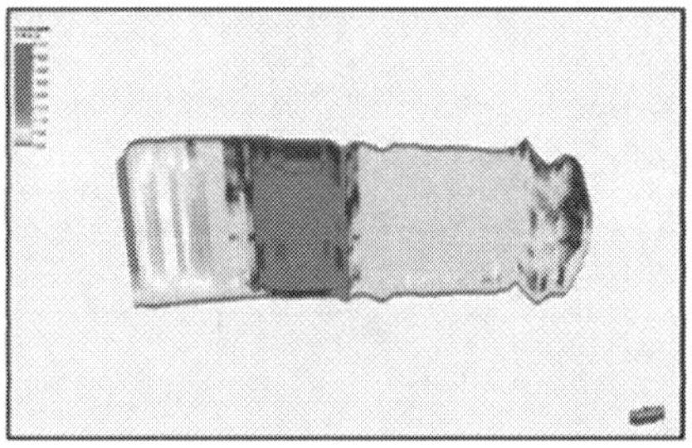

(c) Type III reservoir

Fig. 3 Artificial fracture morphology at optimal displacement for 3 types of reservoirs

4.2 Volume of fracturing fluid

Based on the established fracturing model to simulate the influence of fluid volume, set up 6 groups of fluid volume data, respectively 300, 400, 500, 600, 700, 800 m^3, to analyze and study the fracture length, height, width, as well as support area and transformation volume under the influence of different fluid volumes. From the simulation results, it is found that the fracture length and height as well as the support area and transformation volume show an increasing trend with the increase of fluid volume, and the parameters begin to stabilize when the fluid volume reaches the ranges of 600-700 m^3 (Type I reservoir), 600-700 m^3 (Type II reservoir), and 500-600 m^3 (Type III reservoir). Therefore, 600-700 m^3 of type I reservoir, 600-700 m^3 of type II reservoir and 500-600 m^3 of type III reservoir are considered as the optimal fluid volume. The artificial fracture morphology under the optimal fluid volume values of the 3 types of reservoirs is shown in Fig. 4.

(a) Type I reservoir

(b) Type II reservoir

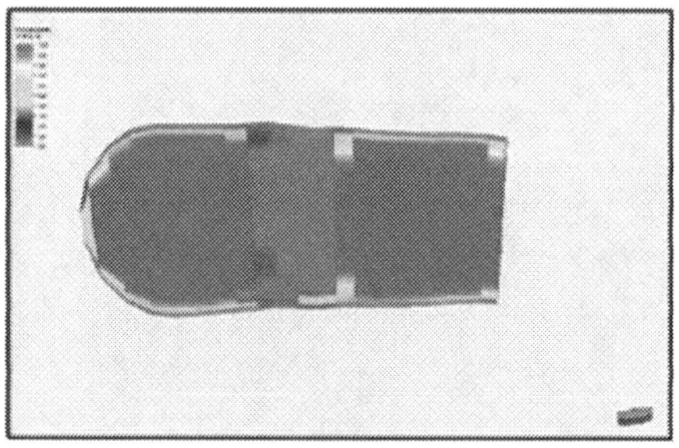

(c) Type III reservoir

Fig. 4 Artificial fracture morphology at optimal fluid volume for 3 types of reservoirs

4.3 Sand proportion

Based on the established fracturing model to simulate the effect of sand proportion, set the sand proportion as 5%, 10%, 15%, 20%, 25%, and 30%, and analyze the fracture length, height, and width, as well as the support area and thetransformation volume of the fracture with different sand proportion. From the analysis of the simulation results, it is concluded that the fracture width increases with the increase of sand proportion, and the fracture length, support area, and transformation volume first increase and then decrease with the increase of sand proportion, and then reach the peak value in the range of 15%-20% (Type I reservoir), 15%-20% (Type II reservoir), and 20%-25% (Type III reservoir), so that the optimal sand proportion values are considered to be 15%-20% for type I reservoir, 15%-20% for type II reservoir, and 20%-25% for type III reservoir. Therefore, it is considered that the optimal sand proportion is 15%-20%

for type I reservoir, 15%-20% for type II reservoir, and 20%-25% for type III reservoir. The artificial fracture patterns under the optimal sand proportion for the 3 types of reservoirs are shown in Fig. 5.

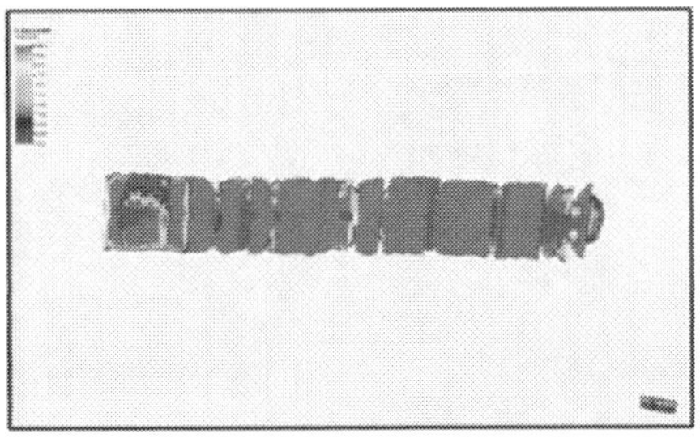

(a) Type I reservoir

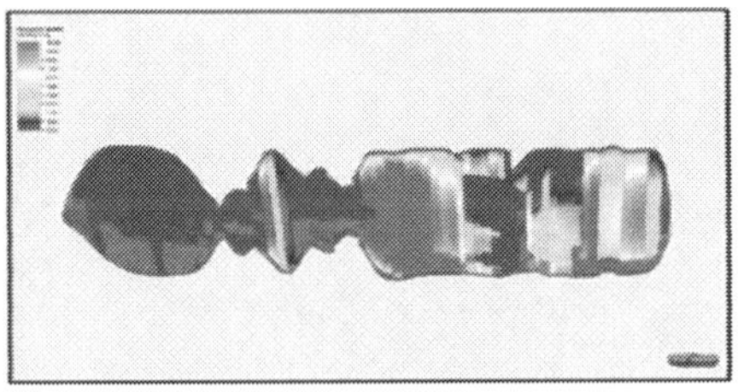

(b) Type II reservoir

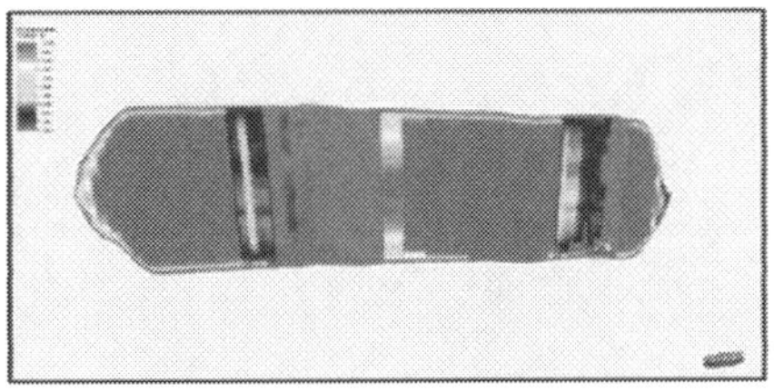

(c) Type III reservoir

Fig. 5 Artificial fracture morphology at optimal sand proportion for 3 types of reservoirs

4.4 Proportion of prepad fluid

Based on the established fracturing model to simulate the effect of the proportion of prepad fluid, set the proportion of prepad fluid as 20%, 30%, 40%, 50%, 60% and 70%, respectively, and analyze the fracture length, height and width as well as the support area and transformation volume of the fracture with different proportions of prepad fluid. From the analysis of the simulation results, it is concluded that the fracture length, height, support area and transformation volume firstly increase and then decrease with the increase of the proportion of prepad fluid, and reach the peak in the range of 50%-60% (Type I reservoir), 40%-50% (Type II reservoir) and 30%-40% (Type III reservoir). Therefore, it is considered that the optimal proportion of prepad fluid is 50%-60% for type I reservoir, 40%-50% for type II reservoir and 30%-40% for type III reservoir. The artificial fracture patterns under the optimal proportion of prepad fluid for the 3 types of reservoirs are shown in Fig. 6.

(a) Type I reservoir

(b) Type II reservoir

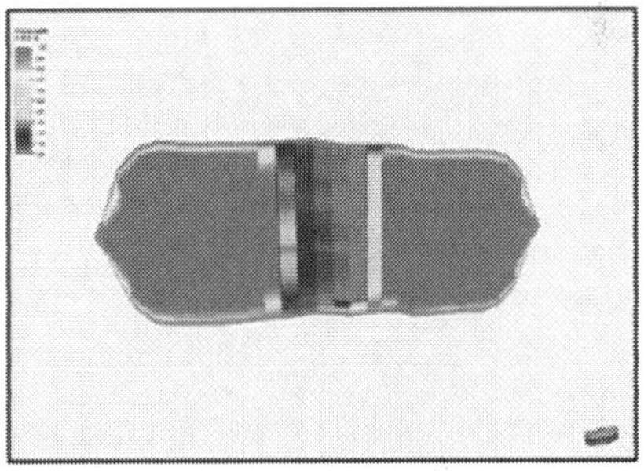

(c) Type III reservoir

Fig. 6 Artificial fracture morphology at optimal proportion of prepad fluid for 3 types of reservoirs

5. Conclusion

(1) Using the 3D original seismic data body to generate the target layer ant-body, the natural fracture model of ant-body of 3 different reservoir types in Block X of Bozi is generated, which provides the geological basis for the integration demonstration;

(2) Comprehensive dynamic and static data and multidisciplinary methods are used to establish a 1D rock mechanics profile model, combine the geological model with the deep rock-forming theory, establish 3D geomechanical models for 3 different reservoir types in Block X of Bozi, and comprehensively analyze and evaluate the matching relationship between the horizontal stresses and the distribution of fault strikes and tensile natural fractures, so as to provide the geomechanical foundation for the assessment of the integrated reservoirs after remodeling;

(3) Comprehensive geological model and geomechanical model, closely combined with engineering parameters, to establish an integrated true 3D artificial fracture model, on the basis of which the optimization of the parameters

of displacement of fracturing fluid, volume of fracturing fluid, sand proportion research and design was completed, the optimal displacement is considered to be 6-7 (m^3/min) for type I reservoir, 5-6 (m^3/min) for type II reservoir, and 4-5 (m^3/min) for type III reservoir; the optimal fluid volume is considered to be 600-700 m^3 of type I reservoir, 600-700 m^3 of type II reservoir and 500-600 m^3 of type III reservoir; the optimal sand proportion is 15%-20% for type I reservoir, 15%-20% for type II reservoir, and 20%-25% for type III reservoir; the optimal proportion of prepad fluid is 50%-60% for type I reservoir, 40%-50% for type II reservoir and 30%-40% for type III reservoir. Provide guidance for field construction.

Reference

1. Guo Honghui; Feng Jianwei; Zhao Libin. Characteristics of passive strike-slip structure and its controlling effect on fracture development in Bozi-Dabei area, Tarim Basin [J]. Oil & Gas Geology, 2019,44(04):962-975.
2. Wang Zhimin; Wang Cuili; Xu Ke; Zhang Hui; Chen Naidong; Deng Hucheng; Hu Xiaofei; Yang Yuyong; Feng Xinluo; Du Yu; Lei Sifan. Development characteristics and controlling factors of tectonic fractures in ultra-deep tight sandstone: A case study of Lower Cretaceous reservoirs in Bozi-Dabei area, Kuqa Depression, Tarim Basin [J]. Natural Gas Geoscience, 2019,34(09):1535-1551.
3. Yang Xuewen; Wang Qinghua; Li Yong; Lv Xiuxiang; Xie Hui Wen; Wu Chao; Wang Cuili; Wang Xiang; Mo Tao; Wang R. Formation mechanism of the Boz-Great North trillion-square atmosphere in the Kuqa foreland thrust belt [J]. Geological front, 2022, 29 (6) : 175-187. The DOI: 10.13745 / j. sf, sf. 2022.8.18
4. Guo Honghui; Feng Jianwei; Zhao Libin. Characteristics of passive strike-slip structure and its controlling effect on fracture development in Bozi-Dabei area, Tarim Basin [J]. Oil & Gas Geology, 2019,44(04):962-975.
5. Huang Peiming. Fracture characteristics and quantitative prediction of ultra-deep tight sandstone reservoirs in Bozi-Dabei zone, Kuqa Depression [D]. China mining university, 2023. DOI: 10.27623 /, dc nki. Gzkyu. 2023.002197
6. Zhang Hui; Ju Wei; Xu Ke; Ningweiaceae; Yin Guoqing; Wang Zhimin; Yu Guodong. Prediction of current in-situ stress of ultra-deep tight sandstone reservoirs in Bozi gas reservoir, Kuqa Depression [J]. Xinjiang Petroleum Geology, 2019,44(02):224-230.
7. Gou Bo, Guo Jianchun. Optimization of large scale fracture parameters based on fine geological model [J]. Oil & Gas Geology,2013(6):809-815.
8. Qu Zhanqing, Huang Desheng, Bruce Lee, et al. Research and application on fracture parameter optimization of fractured horizontal Wells in low permeability gas reservoir [J]. Fault Block Oil & Gas Field,2014(4):486-491.
9. ZHONG Sen. Optimization design of key parameters for staged fracturing of horizontal Wells in SF gas field [J]. Fault Block Oil & Gas Field,2013(4):525-529,534.
10. Carter R.D. Derivation of the general equation for estimating the extent of the fractured area. Drilling and Prod. API 1957, 261-270.

Research on the Assessment Technology of Land Available for PV

Jiaheng Li [2], Mengjin Hu [1], Hongkui Li [2], Mengju Wei [1], Zhao Liu [1], Xuefei Liu [1]

[1] State Grid Hebei Economic Research Institute, HeBei, 050000, China
[2] State Grid Shijiazhuang Electric Power Supply Company, HeBei, 050000, China

Abstract. Based on the high-resolution satellite image data, the information mining technology of the available surface elements of PV is studied, and the investigation of the available surface elements of PV in 98 counties and cities of South Hebei grid is realized. Based on the large-scale and high-resolution remote sensing data obtained by multi-source remote sensing data fusion technology, the depth-learning-based surface feature recognition technology for photovoltaic development is studied. Based on the method of automatic identification and artificial combination of depth-learning, it can identify the available ground elements (roof, water surface, road surface, dry beach, etc.), the available surface elements of PV in 98 counties and cities of Hebei South Grid were obtained. From the overall point of view, the photovoltaic land, the building occupies the main position, in the four cities are relatively high, are in the 6% ~ 15%

Keywords: Photovoltaic, Land type, Assessment of land resources

1. Introduction

By the end of 2017, China had installed 130 gigawatts of photovoltaic (PV) power, but the country's solar resources and electricity consumption are extremely uneven, leading to a persistent problem of absorption, to this end, the state issued a number of targeted policies and measures to encourage the development of distributed power plants in the central and eastern regions. In the Central and eastern regions of China, where land for photovoltaic construction is relatively scarce, how to comprehensively evaluate the potential of photovoltaic power generation in light of solar energy resources, considering the constraints of topography, landforms and the nature of land use, and regarding the government's industrial policy designation, and market investment behavior will have great practical significance.

There have been a number of domestic studies on the potential of photovoltaic power generation. Zhou Yang et al have analysed the spatial and temporal distribution of solar energy resources in Jiangsu province and estimated the amount of rooftop installed capacity in the province, the potential of solar energy resources in Jiangsu province is analyzed in terms of the abundance and stability of solar energy resources, and the spatial and temporal distribution characteristics of solar energy resources in Shanxi province are analyzed by Yan Jiahai, the potential of solar energy resources in Shanxi province is analyzed in terms of solar energy available days, stability and optimal utilization period Liang Yulian analyzed the spatial and temporal distribution characteristics, abundance and stability of solar energy resources in South China, and discussed the suitability of solar energy development under different landforms and underlying surfaces Li Ke et al selected 3 factors of total solar radiation, sunshine duration and effective sunshine days to analyze the land solar energy resource potential in China by using multi-index evaluation method By analyzing the conversion efficiency of monocrystalline silicon modules under different irradiance in Shanghai area, Wang calculated the spatial distribution of PV theoretical annual power generation in China Xie Guohui and others introduced the development potential assessment system of PV power station bases, and Marcel and others used PVGIS data sets to estimate the annual power generation of PV systems in EU member states, the relationship between PV potential and power load in different countries is analyzed. For the potential of photovoltaic power generation, the above research work is involved in different fields, but there are still some problems in the content, such as the lack of pertinence or the lack of comprehensive consideration of factors, so it is very meaningful to explore a comprehensive quantitative evaluation method.Principle of Power Line Communication and Its Application in Energy Internet

2. Investigation of available surface elements of PV

2.1 Calibration and stitching of remote sensing images

This project will use ERDAS remote sensing image processing software, through the geometric correction module of the software, take the State Bureau of Surveying and Mapping data as the control image, and carry out the geometric precision correction of the satellite image, the correction accuracy is less than 1 pixel. After geometric fine correction, the image is a control image, and the registration accuracy is less than 0.5 pixels.

In order to ensure the quality of the corrected image and the convenience of the follow-up processing, the following requirements should be followed in the calibration:

The requirements of the control points: the three-dimensional coordinates of the control points should be measured in the field, the points should be set at the obvious corners, the control points should be distributed evenly, and they should be selected in a stable and unchangeable place. The control point used for orthophoto correction of the reference image and the monitoring image should be the same as far as possible, so as to reduce the deviation of two-phase image collocation;

1) The requirements of the correction model should be consistent with the mathematical model used for two-phase image correction;

2) The selection of resampling mode mainly includes bilinear interpolation and bicubic convolution. The information fidelity in the resampling process is higher by using bicubic convolution than by using bilinear interpolation. With the replacement of the computer, the computing speed is no longer a problem. The practice shows that the operation time of the two resampling modes is almost the same, but the image quality varies greatly, so the bicubic convolution method is usually required in the process of image resampling.

2.2 Research on multi-scale segmentation algorithm of remote sensing image

In this paper, height information and vegetation index are added to the existing multi-scale segmentation methods. In this study, DSM was used to represent height information, while NDVI was the most representative vegetation index. As well as spectral and shape information, height information, as a factor of heterogeneity, is called DSM heterogeneity. The image of NDVI is represented by the weight of NDVI, which is set according to different NDVI, and reflects the image of NDVI in the process of image segmentation. The formula for calculating the algorithm is as follows:

$$f = w_{NDVI} \cdot \left(w_{color} \cdot h_{color} + w_{shape} \cdot h_{shape} + w_{DSM} \cdot h_{DSM}\right)$$

Where w_{NDVI} describes the effect of NDVI; h_{color} , h_{shape} and h_{DSM} for spectral, shape, and DSM heterogeneity, respectively, w_{color}、w_{shape} and w_{DSM} for the weight of spectral, shape, and DSM heterogeneity, respectively. DSM can effectively describe the complex building environment, the building is designed and built, not the natural formation, so it is unexpected and unpredictable. Compared with other types of ground features, the DSM of buildings is obviously different and has clear boundaries. The multi-scale segmentation method not only uses DSM as a layer to calculate the spectral heterogeneity, but also as a factor to determine whether the adjacent objects merge.

$$h_{DSM} = DSM_{MDiff} + \left(Pn_{diff} + PDSM_{diff}/En_{diff}\right)$$

$$DSM_{MDiff} = DSM_{mean1} - DSM_{mean2}$$

DSM_{MDiff} is the difference between the average DSM of adjacent objects before merging, and En_{diff} is the number of adjacent pixels of adjacent objects. Pn_{diff} is the number of pairs of adjacent pixels in which the DSM difference is greater than 3, and $PDSM_{diff}$ is the total value of the DSM difference of adjacent pairs of adjacent pixels in which the DSM difference is greater. DSM_{mean1} and DSM_{mean2} are the average DSM values of adjacent objects, respectively. h_{DSM} is calculated specifically as shown in Figure 10, where two objects 1 and 2 are represented in green and yellow, respectively, and the numbers are the DSM values of the pixels, respectively, and the DSM averages of objects 1 and 2 are calculated, 1.07 and 4.27, respectively, and calculated DSM_{MDiff} as 3.2. The logarithm of the adjacent pixels of the two objects is then Pn_{diff} calculated to be 5 pairs. The number of pairs of adjacent pixels larger than 3 is 5, and the sum of DSM difference of these 5 pairs is 20, i. e. $PDSM_{diff}$ is 20.

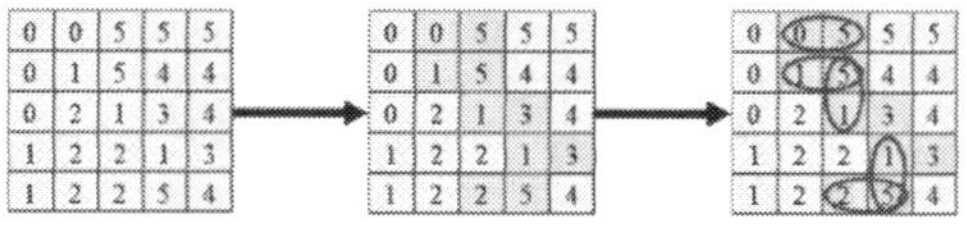

$DSM_{MDiff} = 3.22$ $En_{diff} = 5$ $PDSM_{diff} = 20, Pn_{diff} = 5$

Fig1. hDSM calculation example

w_{NDVI} is determined by the average NDVI of the two adjacent objects before the merge. The formula is as follows:

$$NDVI_{mean} = (NIR_{mean} - R_{mean})/(NIR_{mean} + R_{mean}) \quad (1)$$

$$\begin{cases} w_{NDVI} = 1, NDVI_{mean1} \cdot NDVI_{mean2} > 0 \\ w_{NDVI} = 0, NDVI_{mean1} \cdot NDVI_{mean2} < 0 \end{cases} \quad (2)$$

Where, $NDVI_{mean1}$ and $NDVI_{mean2}$ are the average NDVI of the first two adjacent objects to be merged, respectively.

2.3 Target region extraction

BP neural network is used to extract the target area. BP neural network is a multi-layer feed-forward network. It uses error Backpropagation as learning rules to supervise learning. It trains a large number of condition sets of objects to be classified as known samples, find out the relationship between the feature attributes (patterns) of the input samples and the classification categories. BP neural network has a simple structure and strong Operability. It has been widely used in image classification because of its mature technology and clear processing flow. Based on the selected samples, the BP neural network model is constructed by taking the extracted optimal feature set as the input of BP neural network to realize the recognition of surface elements.

The BP (back propagation) neural network, a multi-layered Feedforward neural network trained by an error back propagation algorithm developed in 1986 by scientists led by Rumelhart and McClelland, is now the most widely used. The aim of BP neural network is to minimize the error mean square deviation between the actual output value and the expected output value of BP neural network. The calculation process of the basic BP algorithm includes forward calculation (i. e. the forward propagation of the signal) and backward calculation (i. e. the backward propagation of the error) . In the forward propagation, the input signal acts on the output node through the hidden layer, and generates the output signal through the nonlinear transformation to judge whether the actual output matches the expected output or not, the error is propagated backward. Error back-propagation is to transfer the output error from the hidden layer to the input layer, and distribute the error to all units in each layer, so as to adjust the weight of each unit. The error decreases along the node gradient when the connection strength between the input node and the hidden node, the connection strength and the threshold value between the hidden node and the output node are adjusted. Then, through multiple training, to obtain the corresponding with the minimum error of the weight and threshold) , stop the training.

The BP neural network classification model is trained according to the input of the sample of the image object selected artificially, the characteristic value of the image object is taken as the input, analyzes and judges whether it is a building, and outputs the result, if the error is large, then adjust the weight value to calculate the output again until the classification precision is satisfied. If the training results meet the requirements, the network model is used to extract the building. The specific training process is as follows:

1) Select a group of training samples, the image object's characteristic is the input information, whether it is the building as the expected output. The input data has M eigenvalues X = (x, x, ... X) ^ t, the input layer has M neurons. The expected output is. Take each feature as each node of the input layer.

2) The input layer passes through each hidden layer in turn, and the output x = γ ^ ((L)) , l = 1,2, ... L, L is the number of implied layers.

3) Reach the output layer, calculate the actual output of the network, and the expected output, the error. At this point, complete the entire network forward propagation.

4) The error of forward propagation calculation is propagated from the output layer, through the middle layer, to the first hidden layer, and distributed to the neurons of each layer according to different weights.

5) Combining with the error of each neuron, the connection weight of each neuron is revised repeatedly.

6) Repeat steps 2 and 3, terminate the training if the error meets the requirement, otherwise continue the iterative training until a satisfactory neural network model is obtained. In order to solve the problem that BP neural network is not easy to converge, the initial weights and thresholds of BP neural network are often initialized randomly. Setting appropriate learning rate and momentum term is also an effective method to improve the convergence rate of network training.

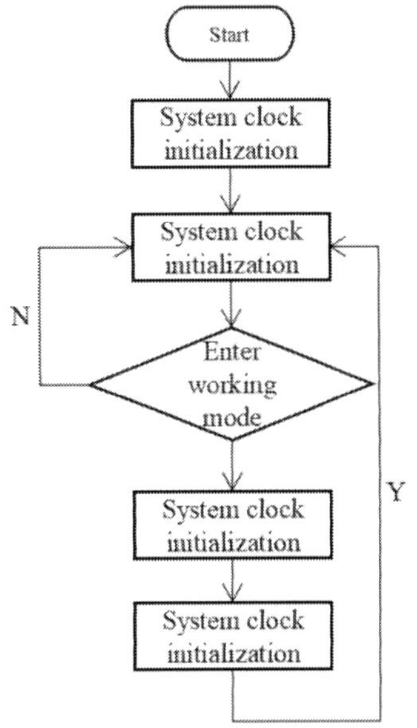

Fig. 2 Algorithm flow

3. Case Study

By analyzing the land-use situation of each city and counting the land resources available for solar energy, it can be seen that the main types of land-use in the cities in southern Hebei are cultivated land, grassland and urban buildings, however, the development and utilization value of cultivated land and grassland is low, and it does not have the condition of photovoltaic development and utilization. On the whole, the area of land available for PV in south Hebei network is 8,530 km^2, including 1,426.27 km^2 in Baoding, 1,220.55 km^2 in Cangzhou, 1,941.84 km^2 in Handan, 559.75 km^2 in Hengshui, 1,866.44 km^2 in Shijiazhuang and 1,514.81 km^2 in Xingtai. Among the four types of photovoltaic usable land, buildings occupy the main position, and are relatively high in the four cities, all in the 6% ~ 15% . The other three types of area, relative to the building, accounting for less, all three of the total area of the city to 1% of the total area. In terms of the area of land available for PV, Shijiazhuang and Handan have the largest area, at about 1,900 km^2, followed by Xingtai and Baoding at about 1,500 km^2, Cangzhou at about 1,220 km^2 and Hengshui at least, only about 560 km^2 of photovoltaic land is available.

Table 1 Land use type in southern Hebei

Cities	Buildings	Bare land	Roads	Dry beaches	Water bodies	Others	Total
Baoding	1295.24	65.20	39.43	26.40	108.68	20701.05	22236
Cangzhou	1143.68	4.31	33.69	38.87	499.38	12394.07	14114
Handan	1856.21	32.93	40.62	12.08	26.19	10087.97	12056
Hengshui	517.96	2.68	25.87	13.24	37.04	8254.21	8851
Shijiazhang	1722.26	72.85	46.26	25.07	94.81	12072.75	14034
Xingtai	1421.17	33.96	41.75	17.93	22.47	10900.72	12438

4. Summary

In order to reduce the cost of PV Land Use Calculation, the project combines object-oriented and deep learning methods. Considering the different types of PV land use, the multi-scale segmentation method is adopted firstly, to segment the remote sensing image and lay a good foundation for the following land use extraction, then in order to give full play to the performance of the classifier, a feature selection method based on feature clustering is proposed, by selecting a small number of samples, the optimal feature set of land use extraction is selected, which contains enough category information. The precision of the buildings extracted with optimal features is similar to that of the buildings extracted with all features, but the buildings extracted with optimal features will reduce the time of calculating a large number of features and be more efficient. BP neural network is used to extract the available land resources of PV by using the optimal feature set, which provides a good data support for PV potential assessment.

Acknowledgments

This research was supported by State Grid Hebei Economic Power Co, Ltd,Investigation on the potential of solar energy and solar energy in provinces based on high-resolution satellite images,Grant B704JY210065.

References

1. Zhou Yang, Wu Wenxiang, Hu Ying, etc.. Assessment of available solar energy resource potential in Jiangsu province. Renewable energy, 28(6) : 10-13.
2. Yan Jiahai, Zhang Dongfeng, an Wei, etc. . Spatial-temporal distribution characteristics and utilization potential of solar energy resources in Shanxi province. Dry weather, 2014,32(5) : 712-718.
3. Liang Yulian, Shen Yanbo, Bai Long, etc.. Assessment and development potential of solar energy resources in South China. Applied Meteorology, 28(4): 481-492.
4. Li Ke, He Fanneng. Regional analysis of the development potential of terrestrial solar energy resources in China. Advances in geographical sciences, 29(9): 1049-1054.
5. Wang Li-zhen, Tan Hong-wei, Zhuang zhi, etc. . Research on the potential of photovoltaics power generation in our country based on GIS platform. Journal of University of Shanghai for Science and Technology, 2014,36(5) : 491-496.

Numerical analysis of influencing factors on wave load for an amphibious aircraft

Xianjiao Gao[1, 2, *], Tengjiao Sang [1,2]

[1] AVIC Special Vehicle research Institute, Hubei Jingmen, China
[2] Key Aviation Scientific and Technological Laboratory of High Speed Hydrodynamic, Hubei Jingmen, China

Abstract. Numerical method of wave load for an amphibious aircraft is carried out in this paper, and its effectiveness is verified by comparing with towing tank model test. It shows that at the same wavelength and speed, the aircraft's wave load is positively correlated with the change of wave height, but with the increase of wavelength, the difference shows a gradually decrease. And wave load is positively correlated with the speed, the greater the speed, the greater the wave load is. The wave load decreases with the increase of the wavelength. The peak load generally occurs in the short wave state of 1.2 times the length of the fuselage. With the increase of the wavelength, the aircraft load gradually decreases.

1. Introduction

In recent years, with the rapid development of mass storage of computer, numerical simulation plays an increasingly important role in study of the wave motion response characteristics of amphibious aircraft. Ma Zenghui used ALE (Arbitrary Lagrange-Euler) method to describe the computational fluid domain, used penalty function coupling method to describe the coupling between aircraft and water body, and studied the motion response of aircraft taking off and landing on different wave surfaces and the dynamic characteristics of fuselage bottom pressure [1]. Chen and others systematically studied the landing performance of amphibious aircraft in waves using RANS equation combined with FVM and VOF methods [2]. Li Xinying carried out numerical simulation research on seakeeping of the hull of an amphibious aircraft based on the sliding grid method, and compared with the model test results to verify the reliability of the method [3].

Based on the STAR-CCM software, this paper uses the Reynolds mean N-S equation and the SST k-w two-equation model to solve the unsteady flow field on the wave surface, the fluid volume fraction method (VOF) is used to capture the free surface, and comparison with the model test results is made to verify the reliability of the numerical simulation method. Based on this, this paper studies the influence of the parameters of taxiing speed, wavelength and wave height on the water skiing load, which has some reference value for the evaluation of the wave load of amphibious aircraft.

2. Numerical calculation method

2.1 Governing equations

Continuity equation and momentum equation are universal equations for numerical solution of viscous fluid motion, and their basic form is:

$$\frac{\partial \rho}{\partial t}+\frac{\partial}{\partial x_i}(\rho u_i)=0 \tag{1}$$

$$\frac{\partial}{\partial t}(\rho u_i)+\frac{\partial}{\partial x_j}(\rho u_i u_j)=-\frac{\partial p}{\partial x_i}+\frac{\partial}{\partial x_j}(\mu\frac{\partial u_i}{x_j}-\rho\overline{u_i^{'}u_j^{'}})+S_i \tag{2}$$

Where $i, j=1,\ 2,\ 3$, μ is dynamic coefficient of viscosity, $u_i^{'}$、$u_j^{'}$ are pulsation values of velocity component, P is the mean value of pressure; S_i is the generalized source type of momentum equation.

Turbulence model should be introduced to enclose above equations. In this paper, the widely used SST $k-\omega$ two-equation model is adopted.

2.2 Boundary conditions and mesh generation

Because amphibious aircraft are symmetrical in structure, and the flow field around the aircraft is also symmetrical about the longitudinal section during navigation. In order to save calculation cost, this paper uses semi-model to carry out the numerical simulation. The computational domain is created by a cuboid with a size of 4λ×2L×2L in length, width and height, respectively, whereλ is wavelength, L_H is the hull form length of the seaplane.

* Corresponding author: 2495648607@qq.com

In order to capture the motion attitude of the aircraft as well as to reduce the computational cost, a trapezoidal grid layout is introduced to conduct different degrees of mesh transition and encryption for the entire computational domain. The area in front of the fuselage head consists of wave generating zone of 0.8λ and encounter zone of 0.2λ, area behind the rear part of the seaplane is composed of wake zone of 0.7λ, transition zone of 0.3λ, and absorbing zone of 1λ. Grid distribution of symmetry plane and the seaplane are shown in Figure 1 and figure 2, respectively.

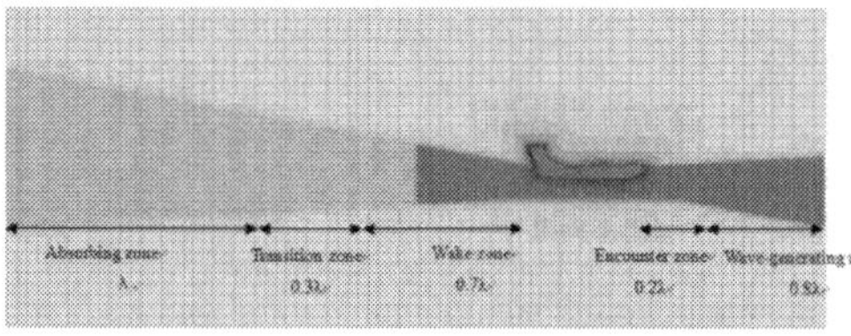

Figure 1. Mesh generation of computational domain.

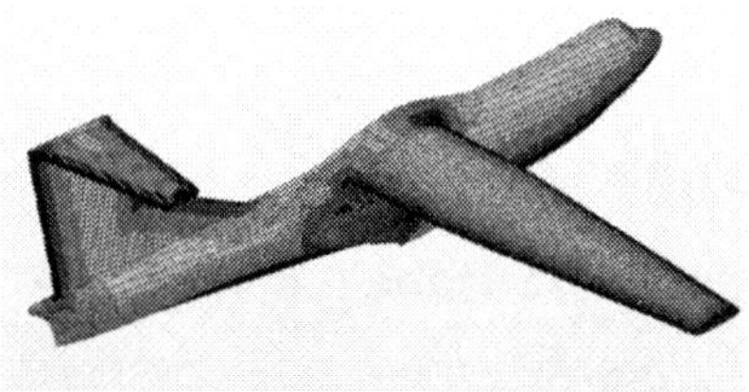

Figure 2. Mesh distribution of seaplane.

The $k-\omega$ shear stress transport (SST) turbulence model is adopted to solve the stress tensor based on the research of turbulence model effect in numerical wave simulation. The second-order upwind scheme is used to solve convection term, free surface of the wave is captured by the volume of fluid domain (VOF method, and the SIMPLE algorithm is applied to address the pressure-velocity coupling.

2.3 Validation of numerical method

In order to validate the accuracy of numerical method, simulation of typical wave state is made by using the STAR CCM+ commercial software package, and comparisons with towing tank model test are made. Figure 3 and Figure 4 shows comparison of gravity load in time and wave pattern, respectively. It can be seen that the time-history curve of gravity load obtained by numerical simulation is basically consistent with the model test result.

It can be seen in Figure 4 that the CFD method can well capture the cavitation near the fuselage step and the splash at the rear bilge line, but it is difficult to capture the large deformation of the free surface such as flap splash. This is because the numerical calculation is implemented based on grid technology. In the grid of the computational domain, each point has a fixed number of predefined adjacent points 4. The connection relationship with the adjacent points is used to solve the grid control equation and transfer the solution information to the entire computational domain. When the grid deformation becomes large to a certain extent, the connection relationship between the grids is difficult to maintain if no error is introduced, although the grid can be subdivided in the simulation. However, it will eventually lead to an increase in the accumulation of errors, which will make it impossible to capture such large-scale deformation.

In terms of numerical simulation accuracy, the numerical load simulation value is 1.78g, with a deviation - 10.1% compared with the model test value of 1.98g, which is within the acceptable range of the engineering application of the numerical simulation of seaplane, indicating that the CFD method is feasible for the study of wave load for the seaplane.

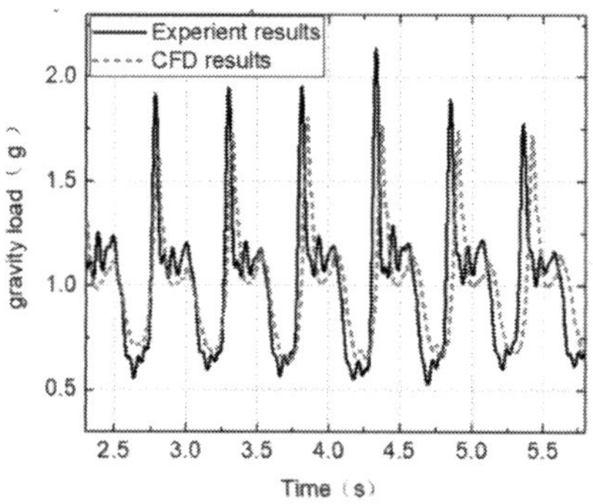

Figure 3. Comparison of time-history curve of wave load between simulation and experimental result

(a) experiment wave pattern

(b) CFD wave patte rn

Figure 4. Comparison of wave pattern between calculation and experiment

3. Analysis of influencing factors of wave load

Calculations of conditions listed in Table 1 are performed to study the effect of wavelength, wave height and sailing velocity on gravity load of the seaplane.

Table 1. Research conditions.

Wave height /hull length	Wave length /hull length	Sliding speed /takeoff speed
0.02、0.03、0.04	1.2、1.8、2.4、3.0、4.3	0.28、0.42、0.56、0.70、0.85

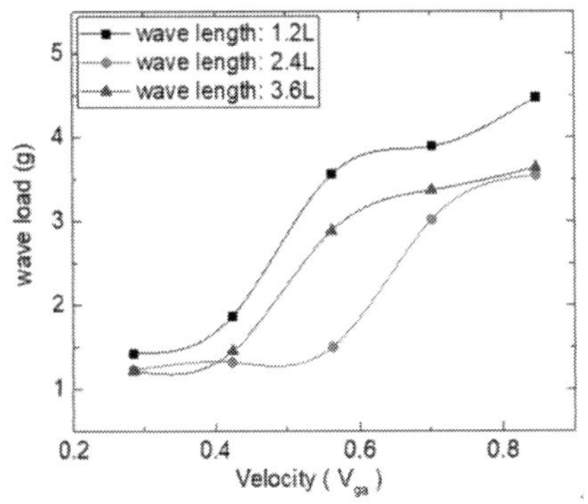

Figure 5. Influence of velocity on wave load (wave height:0.03L)

Figure 5 shows the variation curve of wave load at the center of gravity with speed at fixed wave height and different wavelengths. It can be seen that the load is gradually increasing with the increase of speed, but the impact of speed variation on load is significantly stronger in the transition stage than the navigation and taxiing stage, because during the transition stage, the taxiing area of the rear body is gradually reduced with the increase of speed, and this leads the seaplane into a single step taxiing stage [5]. And change of speed has a significant impact on the aircraft attitude, there, the impact load generated by the collision force between the fuselage and wave surface changes obviously with speed.

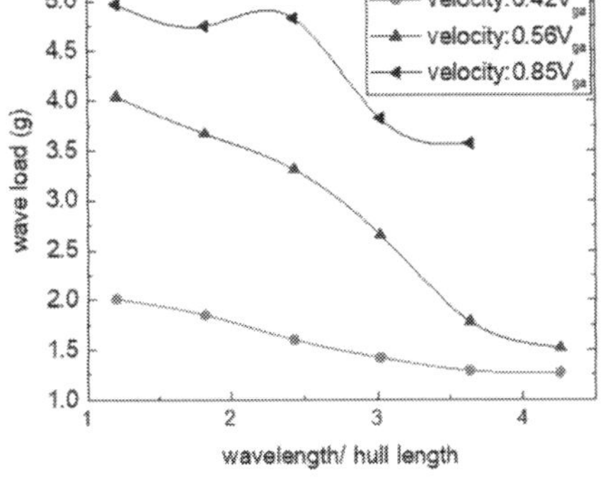

Figure 6. Influence of wave length on wave load (wave height:0.04L)

Figure 6 shows the wavelength effect on wave load performance. It can be seen that as the wavelength increases from 1.2L to 4.3L, the wave load generally shows an opposite feature against the wavelength. Maximum load of the aircraft at fixed speed and wave height occurs at the short wave length of 1.2 times the length of fuselage. This is because the aircraft has a phenomenon of wave hopping, that is, after the aircraft jumps off the water surface in a short wave state, it will not collide with the water surface until it has flown over 1-2 wave peaks, and the fuselage will suffer a large impact force, which caused a relatively large impact load.

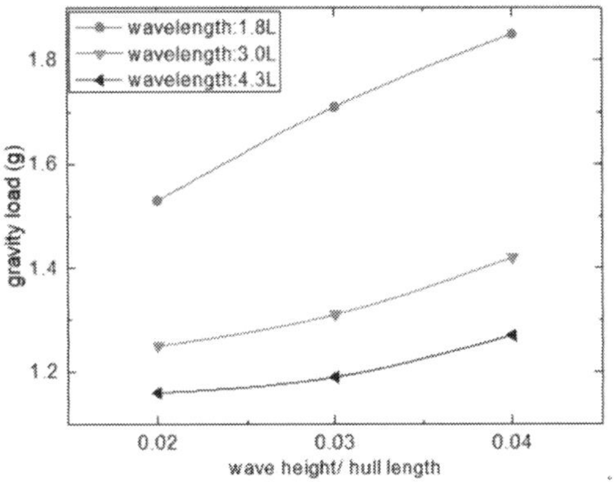

Figure 7. Influence of wave height on wave load(velocity:0.42Vga)

Figure 7 shows variation of load with wave height, it is obviously that the load is positively influenced by wave height. However, the impact of wave height variation on load in short wave state is significantly greater than that in the long wave state, so the aircraft should avoid taking off and landing in the short wave state.

4. Conclusions

In this study, numerical simulation method is used to carry out the numerical calculation of wave load for an amphibious aircraft, and the feasibility of the method is verified by comparing with the towing tank model test results. Based on the method, the effects of wavelength, wave height and velocity on wave load are studied, and the conclusions are as follows:

(1) The load value and wave pattern obtained by numerical simulation are in good agreement with the experimental results. The calculation error is controlled at about 10%, which is within the allowable range of engineering application.

(2) Wave load of the aircraft increases with wave height and velocity. Higher sea conditions and high navigation speed will cause the aircraft to bear greater overload.

(3) Wave load is negatively correlated with wavelength. The maximum load occurs at the short wave of 1.2 times the length of the fuselage hull state. When the wavelength reaches to 4 times the length of the fuselage hull, the load gradient caused by the wave height and velocity gradually decreases.

Acknowledgments

This work was financially supported by Special Research on Civil Aircraft Fund of China (No. MJZ5-3N21-5, MJZ5-3N21-1).

References

1. MA Z H, LIU L S, YUE Z, et al. Numerical investigation on seakeeping performance of amphibious aircraft landing on waves [J]. Chinese Journal of Computational Mechanics, 2018, 35(3) : 380-386. (in Chinese).
2. CHEN J C, XIAO T H, WANG M et al. Numerical Study of Wave Effect on Aircraft Water-Landing Performance [J]. Appl. Sci., 2022, 12, 2561.
3. LI X Y, WU B, JIANG R. Numerical simulation of seakeeping performance of amphibious aircraft hull based on sliding mesh method [J]. Ship & Ocean Engineering, 2020, 49(03) : 14-18. (in Chinese).
4. QIN X G, LIU P Q, QU Q L. Numerical simulation on aerodynamics of airfoil flying over wavy water surface [J]. Journal of Beijing University of Aeronautics and Astronautics, 2011, 37(3) : 295-299. (in Chinese) Paterson J. H. Recent development in the hydrodynamic design of flying boats[J]. Journal of Royal Aeronautical Society, 1955, 59: 349-355.
5. HUANG M, CHU L T, LI C H, et al. Seakeeping performance research of large amphibious aircraft [J]. Acta Aeronautica et Astronautica Sinica, 2019, 40(1) : 522-535. (in Chinese).

Analysis of China's Marine Renewable Energy Current Situation

Jian Li, Xiaokun Ma, Xiangnan Wang, Hongming Qiu, Changlei Ma

National Ocean Technology Center, Tianjin 300112, China

Abstract. China's Marine renewable energy is in a period of rapid development. Since the establishment of the national special fund for Marine renewable energy in 2010, a large number of demonstration projects, research and test projects have been strongly supported and promoted, forming a number of technology and equipment with independent intellectual property rights. But as a new energy field, some problems have been found in the process of project implementation, this paper is based on the 2010-2020 projects of marine energy special project research, analyzes the progress of the wave energy, tidal energy, classifies and summarizes the research and test projects, expounds the current marine energy special project in the process of implementation problems and deficiencies, put forward some suggestions for the development of Marine renewable energy special projects in the future.

Keywords: Marine renewable energy; tidal energy, wave energy;

1. Preface

In June 2010, the State Oceanic Administration and the Ministry of Finance jointly issued the Interim Measures for the Management of Marine Renewable Energy Special Funds, which funded the development of domestic Marine energy utilization and Marine energy technology through a special fund of 200 million yuan per year. The first batch of special fund projects have been launched in 2010. According to the schedule, most of the projects have enter the final acceptance stage [1]. During this period, the Marine renewable energy special project was orderly developing, formed a number of independent intellectual property rights of Marine energy transformation and utilization technology, it is also cultivates a group of special talents in the field of Marine renewable energy, and fills in the gaps of the Marine renewable energy in China, especially in the field of Marine power generation. Through support the national finance continuously, now it is has been formed a standardized theoretical research system of Marine renewable energy, but in the practical application of some key technologies it is still greats shortcomings. This paper analyzes the current development of Marine renewable energy from the three parts of research and test for tidal energy, wave energy and demonstration project, and puts forward some suggestions.

2. The analysis for tidal energy power generation research and test

The research and test for tidal energy is a research or test that converted the tidal current energy into electric energy or other available resources. The energy density of tidal current energy is very large [2], which has the incomparable sustainability, regularity and predictability compared with other ocean energy. At present, China's tidal power generation devices according to different needs can adopt towed type, floating type and base type. Among them, the towed type and floating tidal power generation devices have simple structure and strong maintainability, most of which supply power to buoy, latent mark or sea bed base [3]. Because the fixed structure of the towed type and floating type power generation device is limited, the power generation device will not only by the effect of tide, but also may be under the action of the ocean surface waves, so the spindle of power generation device will be extremely complex and easily broken, and it's reliability index is the main factors for the development of towed type and floating type power generation device. In contrast, the base type tidal energy power generation device avoids the adverse factors on the ocean surface, and the power generation efficiency and reliability of the device are relatively improved. If the guide cover or variable pitch technology is used, the power generation efficiency and reliability will be further improved [4]. For example, in August 2012, Harbin Engineering University launched "Haiming I" tidal energy power generation device in Guishan Waterway. The designed power generation of the device was 2150kW, and the diameter of its turbine blade was 4 meters. In June of the following year, "Haiming" was launched in the sea area of Zhaitang Island in Qingdao. The blade diameter of its turbine was 12 meters, and the energy conversion efficiency was about 34%. In 2013, the 20kW the base type tidal energy power generation developed by Northeast Demonstration University was tested at low velocity in the waters of Zhaitang Island in Qingdao. In 2018, the 300kW semi-direct drive horizontal shaft variable propeller unit jointly developed by Guodian United Power and Zhejiang University went to sea test in Ruoshan Island of Zhejiang University and successfully generated electricity. In terms of the grid connection for tidal power generation device, Zhejiang Zhoushan Joint kinetic Energy New Energy Development Company Limited developed LHD modular tidal energy power generating assembling unit, which completed offshore installation in 2016 and completed grid connection to power generation in August. By 2019, it has generated 1.5 million kWh. Although our country tide energy power generation device is mature relatively, but there are many deficiencies in safety maintenance, such as base type tide energy power generation devices are arranged more than in ten meters or dozens of meters deep underwater, its maintainability maybe become the key to limiting tide energy power generation device, because the Marine environment is complicated, especially in the marine energy power generation device， there are some deficiencies in corrosion prevention, mechanical structure and the key component of stiffness, strength, etc. So it needs to be tested for long-term normal operation characteristics of marine energy power generation device. In the terms of efficiency of power generation, different parameter input will make the power efficiency calculation different, and at the present the marine energy industry is not specialized for the scientific and authoritative explanation of the marine energy power generation efficiency. So each project has its own power generation device conversion efficiency calculation formula, which led to the power generation efficiency data matches the requirements of the task, but the generating power is far away from the task.

3. The analysis for wave energy power generation research and test

Wave energy is the most widely distributed energy in the marine. Wave energy power generation device is relatively small but the application range is large. The commonly used wave energy power generation device mainly consists of oscillating water column type, oscillating floating type, pendulum type, duck type, etc. The wave energy power generation device generally works on the surface of the ocean [5], where the sea conditions has a certain impact for the device. Especially for the wave energy power generation device which is without fixed support structure, and its survival ability is the key factor for the power generation device can reach the technical maturity whether or not. Wave power generation technology in our country is relatively mature, for example the wave energy power generation device which is developed by Chinese Academy of Sciences Guangzhou energy has generated more than ten thousand degrees electric power in Wanshan Island in 2017, Shandong university and Jinan far new energy development company joint development the point absorption hydraulic wave energy power generation system which is completed testing in 2016, Ocean university of China developed oscillation float wave energy power generation device and tested in Qingdao zhaitang island sea area successfully in 2014[6]. In general, the wave energy research in our country although started slowly, but the development is relatively rapid, it has made certain achievements which is basic close to the international advanced level in the theoretical research and equipment manufacturing. Recently a horizontal axis rotor turbine wave energy power system adopts the principle of wave energy reciprocating utilization, greatly improve the energy conversion efficiency, but there are more different opinions for the conversion efficiency calculation method for this device. The distribution of the wave energy is average and the energy is easy to obtain, it can solve the problem that regularly replaced the battery for the buoy, submersible ocean observation equipment, and reduce the ocean observation equipment maintenance cycle by using the wave energy. Although wave energy resources in our country and abroad have been certain development, but through the situation of the specific test, it is also exposed some problems, such as a single wave, single current laboratory can analyze the wave energy power device ideal model, but when faces the marine environment, especially the actual environment of ocean waves, it needs consider the efficiency of the surface wind field and the surface flow.

4. The analysis for ocean thermal energy

Ocean thermal energy is mainly in the form of the temperature difference between the surface layer and the bottom layer of seawater stored in the ocean thermal energy. The main source of ocean thermal energy is solar thermal radiation, the ocean receives about 15% of the sun's energy as thermal energy stored in the ocean, the sea surface to 60 m or so of the surface layer of the general temperature of 25-27 ℃, 60-300 m of the middle layer of seawater due to the weakening of the sun's radiation, seawater temperature with the depth of the drastic changes in the depth of the deeper seawater below 300 m due to the lack of sunlight all year round, the temperature The temperature of the deep sea water below 300 m is reduced to about 4℃ due to the lack of sunlight all year round. Generally speaking, the temperature difference of seawater should have a minimum difference of 20℃ in order to generate electricity stably, and low latitude areas such as Cuba, Indonesia, and the South China Sea are ideal places for the use of ocean thermal energy. In addition to the application of ocean thermal energy in the field of power generation, it also has the desalination of seawater, seawater farming and hydrogen production and other aspects of use.

The concept of ocean thermal energy was put forward more than a century ago, but only in the last 40 years to obtain the technical progress. 1881 French physicist Assomval in the world for the first time put forward the idea of ocean thermal energy power generation; in 1929 his student French engineer Claude confirmed the possibility of ocean thermal energy power generation and, the following year, in Cuba, established the world's first ocean thermal energy power station, obtained 10 kW of electricity; in 1979 the United States in Hawaii, the power station of ocean thermal energy power generation, obtained 10 kW of electricity. In 1979, the United States in the waters off the island of Hawaii established the first practical application of ocean thermal energy generation device, the device uses ammonia as a working medium, the power generation power of 50 kW; 1980 Taiwan Power Company had planned to release the heat of the nuclear power plant and the ocean ocean thermal energy power generation and with the selected Hualien, Taitung and other bottom to carry out a joint study; in 1985, China's Academy of Sciences, Guangzhou, China energy Guangzhou Energy Institute of the Chinese Academy of Sciences in 1985, the use of "droplet lifting cycle" method of research, through theoretical calculations when the temperature is lowered from 20 ℃ to 7 ℃ when the release of seawater heat can be lifted to a height of 125 m seawater, and then through the turbine to generate electricity, in 1989, the Guangzhou Energy Institute completed the "droplet lifting cycle" test bed, the installation of a new generation of energy. In 1989, Guangzhou Energy Research Institute completed the "droplet lifting cycle" test bed, with an installed capacity of 10 W and 60 W. In 1993, the United States established an open cycle test device in Hawaii, with a power generation capacity of 210 kW, and the actual output power of 40 to 50 kW, and can produce fresh water; in 2004-2005, Tianjin University completed the research on the use of hybrid ocean energy ocean thermal energy, and 200 W ammonia saturated steam turbine for miniaturisation test was developed.

5. The analysis for ocean salinity gradient energy

Ocean salinity gradient energy is the energy generated by using the potential energy between seawater and freshwater. The principle is that when two different concentrations of salt solutions are mixed, salt ions with high salt concentration will move to the solution with low salt concentration until the concentration of the two solutions is the same, and the ocean salinity gradient energy is to use the chemical potential difference generated by this characteristic and convert it into electricity to generate electricity. At present, the utilisation of ocean salinity gradient energy is still in the laboratory research stage, and there is still a long way to go before offshore demonstration and grid-connected power generation for practical application.

The use of ocean salinity gradient energy relative to other ocean energy use started later, in which the United States, Israel in the ocean salinity gradient energy research is in the leading position, China and Sweden, Japan and other places in recent years have also carried out the corresponding research. In the 1970s and 1980s, Israel and the United States conducted experimental research on hydraulic towers and strong osmosis systems, and the 150 kW ocean salinity gradient energy power generation device built in Israel is currently the largest installed capacity of ocean salinity gradient energy devices; in 1985, the Institute of Fluid Mechanics of Xi'an Institute of Metallurgical Architecture adopted the semi-permeable membrane principle of the research and development of the ocean salinity gradient energy power generation device produces only 0.9-1.2 W (membrane area 14 square metres); China's ocean affairs and ocean affairs and ocean affairs and ocean affairs and ocean affairs and ocean affairs and ocean affairs are the first to use ocean salinity gradient energy in China. In 1985, the Institute of Fluid Mechanics of the Xi'an Institute of Metallurgical Architecture adopted the semi-permeable membrane principle to develop a ocean salinity gradient energy power generation device, which produced only 0.9-1.2 W (membrane area of 14 square metres).

6. The analysis for Marine renewable energy demonstration project

The Marine energy demonstration project is a integration for Marine energy and wind energy and other renewable energy, which organize a power system project based on renewable energy. Most of project use the mature Marine energy power generation device[7], and its stability has been verified in research tests. Marine energy independent power system demonstration project is the focus of the marine energy special project, and the process is in the exploration, although the power generation device which used in the demonstration project is relatively mature. but because the demonstration project is a comprehensive utilization for renewable energy, so the influence for island sea area and the large demonstration project, the structure of the marine energy power generation device, are all the restricting factors of the demonstration project: (1) The using application of the sea area. The application of sea area use is the key element that restricts the "landing of demonstration projects". The planned sea area in the implementation plan of multiple projects is not consistent with the actual using of the sea area, so it results the actual hydrodynamic parameters of the sea area cannot match the requirements of the power generation device. The project team should communicate with the local government in detail before the project is subscribed to ensure that the sea area application can be completed smoothly. The local government should also relax the conditions of the using of the sea area, change the thinking of "political performance first", and cooperate with the Marine energy project; (2) The ecological protection of island. The ecological protection of Island is also one of the factors which should be considered in the Marine energy demonstration project, especially when the implementation plan is compiling, it needs to be fully investigated and analyzed for the impact of the demonstration project on the local island ecology, and protect the local ecological conditions.(3) Standard support. At present, there have been many complementary studies in the field of wind power and photovoltaic, such as GB/T 19115-2018 wind-solar-complementary power generation system, GB/T 29544-2013 off-grid wind-solar-complementary power generation system-safety requirements, etc., but none of them involves the field of Marine energy, so the standards which support Marine energy, wind and light complementary are still blank.

7. The conclusions and recommendations

The tidal energy and wave energy have the characteristics of capturing energy simple, distribution energy wide, and designing the power generation device mature, which have become the hot spot of the current Marine renewable energy utilization. The marine energy utilization in China is still in the development stage, a number of key technologies are still being explored and studied, so this paper put forward the following Suggestions for the future Marine renewable energy development: (1) marine energy projects have analyzed the theoretical data perfectly, but the actual test experience, especially in the reliability of the power generation device, maintainability, environmental adaptability, have no detailed about solution and certification scheme. Therefore, it is necessary to control the design and manufacturing links strictly, strengthen the sea testing, the outline testing and relevant standards should be formulated for Marine r energy power generation devices to promote the development of Marine energy.(2) Establish the project node control mechanism, it needs to inspect the actual process of the project according to the progress of the project regular or not, solve the problems which are found in the project implementation process timely, and avoid to reducing the technical input for the Marine energy power generation device in order to the delay of the projects.(3) Continue to strengthen the construction of the comprehensive support platform for Marine energy, it needs to promote the process of the technology and industrialization of the Marine energy, and provide technical support for the development of China's Marine energy industry.(4) Cooperate with the local government, it needs to negotiate and communicate in the using of the sea areas and ecological protection, give feedback on problems timely, and ensure the Marine renewable energy projects smooth implemented smoothly.

References

1. Compilation of special fund management documents for Marine Renewable Energy [M]. Department of Science and Technology, SOa, 2012:1-308.
2. Lu Wei. Marine energy resources analysis method and reserve assessment [M]. Beijing: Ocean Publishing House, 2009:50-75.
3. Zhang Song. Investigation and Evaluation of offshore wave energy resources in China [J]. Marine Technology, 2013.2:79-81.
4. Wang Ning. International Electrotechnical Commission Renewable Energy Equipment Standard Certification System [J]. Quality and Certification 2017 (06)
5. Zhang Wenjing, Xu Chunhong, Yu Mingru. Research on the construction of standard system for Marine Energy exploitation and utilization in China [J]. Standard Science. 2015(08)
6. Wang Baosen, Xu Chunhong, Chen Hua. The enlightenment of the development and utilization of Marine renewable energy in the world to China [J]. Marine development and management. 2014,31(06)
7. Xu Hongrui, Ma Yue, Li Jing. Thoughts on the establishment of a standard system for wave power generation device [J]. Standard Science. 2014(05)

Current Situation and Prospect in Urban Sewage Treatment Plants

Yao Ding*, Ning Li, Maolin He, Rongxin Zhu and Yin Wang

Southwest Municipal Engineering Design & Research Institute of China, Chengdu 610084, China

Abstract. Sewage is the carrier of resources and energy, which contains great chemical energy and heat energy. The traditional sewage treatment process consumes energy, and the energy consumption is high. Under the background of carbon neutrality, China's urban sewage treatment plants have achieved carbon neutrality operation in terms of energy self-sufficiency and reduction of greenhouse gas emissions. To explore the feasibility of carbon neutrality in sewage treatment plants, the cases of carbon neutrality in sewage treatment plants at domestic and abroad were analyzed. And the implementation approaches and main emission reduction processes to achieve carbon neutrality were summarized. The technical challenges and research directions of future sewage treatment plants in China from the perspective of carbon neutrality are proposed.

Keywords: Carbon neutrality; sewage treatment plant; low-consumption; energy recovery; carbon emission.

1. Introduction

Since the 21st century, the problem of global warming has become increasingly serious and urgent. In 2020, China promised to reach the carbon peak in 2030 and achieve carbon neutrality in 2060. The carbon emissions of the sewage treatment industry account for 1 % ~ 2 % of China's total carbon emissions [1]. Therefore, the carbon emissions generated during the sewage treatment process cannot be ignored. According to statistics, as of 2019, China's urban sewage production was $5.546 \times 10^{10} m^3$, sewage treatment energy consumption accounts for about 0.73 % of the total social electricity consumption, and greenhouse gas emissions reached 91 million tons of carbon dioxide equivalent (2014) [2,3]. Under the background of national carbon neutrality, the green and low-carbon operation is the transformation direction of urban sewage treatment. Generally speaking, the carbon-neutral operation of urban sewage treatment plants has a specific definition, namely energy balance: recovering the energy in sewage to feed back the energy consumption of urban sewage treatment plants to achieve energy self-sufficiency, while reducing the life cycle of sewage treatment plants to make greenhouse gas emissions achieve carbon neutral operation.

To promote the development of carbon neutralization technology in sewage treatment plants, this paper summarizes the latest progress of carbon neutralization technology in sewage treatment plants through the interpretation and analysis of carbon neutralization cases in five blow treatment plants at domestic and abroad. The aim is to provide a reference for the realization of carbon neutralization operations in urban sewage treatment plants in China.

2. Present Situation of Carbon Neutralization in Sewage Plants

2.1 Carbon neutralization status of foreign wastewater treatment plants

Global warming climate crisis is imminent, and countries around the world for this. Based on the goal of carbon neutrality, industries minimize environmental loads while reducing energy consumption. The need for sustainable development is driving new developments in wastewater treatment. These developments are driven by two forces: the improvement of the process flow and the recovery of resources [4]. Achieving carbon neutrality in sewage plants is the goal of the sustainable development of sewage plants in the future [5]. Some countries in the world have been at the forefront of carbon neutrality, and have successively issued a road map to achieve carbon neutrality in sewage plants, providing the advanced experience for the sustainable development of sewage treatment in the world. The Sheboygan wastewater treatment plant in the United States has realized the self-sufficiency of energy in the treatment process through the operation strategy of open source and throttling. As a model of the operation of the wastewater treatment plant in the United States, the plant has achieved a ratio of electricity production to power consumption exceeding 90 %, and a ratio of heat production to heat consumption

* Corresponding author: dingyao1220@foxmail.com

exceeding 85 % [6]. The Netherlands defines carbon neutrality in wastewater treatment plants as a NEWs framework. Sustainable wastewater treatment is nutrient, energy, and reclaimed water, essentially [7]. Based on this concept, the Netherlands will gradually achieve the goal of carbon neutrality. The Steinhof wastewater treatment plant in Germany makes full use of nitrogen, phosphorus, and other resources in wastewater and reuses the treated water resources and sludge. The total exogenous CO_2 of the plant is reduced by 114 %, which can exceed the carbon neutralization target [8]. O.NOWAK et al. [9] analyzed the Wolfgangsee-Ischl wastewater treatment plant and Strass wastewater treatment plant in Austria. The results show that the power generation capacity of the wastewater treatment plant can reach 180 % compared with the energy demand, and the sewage was converted into cooling water as the heat source of the district heating heat pump, which could save up to 45 % of the electricity.

2.2 Carbon neutralization status of domestic wastewater treatment plants

To cope with the problems of water resources and carbon emissions in the future, some foreign wastewater treatment plants are at the forefront of the world in terms of resource recovery and energy utilization. They provide valuable experience for domestic sewage plants to achieve carbon neutrality. Advanced cases of sewage plants in the United States, the Netherlands and Germany, and other countries show that it is feasible to achieve carbon neutrality in sewage plants, and China should also adapt to local conditions. Limited by the technical level of the development of the water industry and many other factors, China has not yet built a truly carbon-neutral sewage plant. Compared with developed countries such as the United States, China's future-oriented concept of sustainable sewage treatment plants was proposed late. In 2014, Academician Qu proposed 'building a future-oriented Chinese sewage treatment concept plant'. In the future, the operational objectives of sewage treatment plants will change from pollutant reduction to water resource reuse, resource and energy recovery, and water ecological restoration [10]. Resource-oriented sewage treatment has gradually become the theme of the future pursuit of the global water treatment industry. China will also actively explore the development path of sewage plants. In the future, China's new concept of sewage treatment plants will achieve the goal of carbon neutrality based on water purification. Based on this concept, some sewage plants in China have begun to explore new modes of sewage treatment. Henan Shangqiu Suixian Sewage Treatment Plant can be regarded as the 1.0 version of the future sewage treatment plant in China. Jiangsu Yixing New Concept Sewage Treatment Plant further builds the sewage treatment plant as a resource plant on version 1.0, opens a new construction concept, and builds the traditional sewage treatment plant into a sustainable development sewage treatment plant for the future.

3. Process development of carbon neutral operation in urban sewage treatment plants

The organic matter in urban sewage is an energetic substance, and its chemical oxygen demand contains the lowest energy of 13-14 kJ / g. In the traditional sewage treatment process, most of the energy is directly mineralized into CO_2 and lost. If captured and converted, it is expected to offset energy input and even external use of production capacity [11]. The key to achieving carbon-neutral operation of municipal wastewater treatment plants is low energy consumption operation and energy recovery, while traditional treatment processes (aerobic aeration, nitrification-denitrification, etc.) are difficult to achieve, and new processes need to be developed. Researchers generally believe that energy can be recycled through 'carbon capture' technology supplemented by anaerobic digestion-cogeneration, heat recovery, and other technologies [12]. At the same time, the use of technologies such as autotrophic nitrogen removal and denitrifying phosphorus removal can reduce energy consumption, thereby achieving carbon neutrality in sewage treatment plants [13].

3.1 Carbon capture technology

The 'carbon capture' technology based on sewage treatment refers to capturing COD in the influent into the sludge and generating energy through anaerobic digestion of the sludge. It is generally to increase the COD capture in the primary treatment of sewage treatment plants to increase energy production [14]. 'Carbon capture ' technology is mainly realized by high rate activated sludge process(HRAS) [15] and chemically enhanced primary treatment process(CEPT) [16].

(1) HRAS

Because of its short hydraulic retention time(HRT), sludge retention time(SRT), and high sludge load, HRAS mainly removes organic matter by flocculation and adsorption. It can adsorb and retain the organic matter in the water to the maximum extent and then become the first selected process of 'carbon capture' technology [17](Figure 1). Chen [18] conducted experiments and found that maintaining a low SRT (0.8d) and HRT (1.5h), using HRAS technology. Guven et al. [19] conducted a pilot study on the influent of the sewage treatment plant, and used different HRTs to run the HRAS process. It was found that under the condition of HRT = 60min, the total COD removal rate reached the highest 59 % while minimizing SRT (0.35d), the mineralization rate of adsorbed COD was the lowest (23 %), which greatly improved the 'carbon capture' efficiency. A large number of studies have shown that shorter HRT and SRT are the key to achieving high carbon capture and low biomineralization in HRAS. A pilot study was also carried out in a sewage treatment plant. The COD capture rate of HRAS process reached 87 %, and the COD mineralization rate was only 16 %. Compared with the traditional activated sludge process, the COD recovery rate was greatly improved[20]. In the context of carbon neutrality, HRAS has attracted attention due to its high carbon capture efficiency and low mineralization rate. Based on short HRT and SRT, HRAS can achieve the efficient capacity of back-end energy recovery. At the same time, it can be used in urban sewage treatment plants to significantly reduce the floor area of structures and investment and operation costs, highlighting the advantages of HRAS in urban sewage treatment plants.

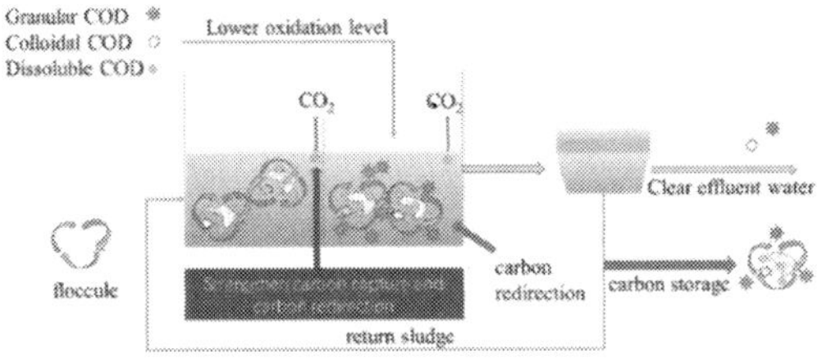

Figure 1 HRAS

(2) CEPT

CEPT removes organic matter in the influent by chemical coagulation, the same as. It was found that CEPT removed more than 60 % of particulate organic matte r[21]. With different coagulants and flocculants, the capture efficiency of CEPT on COD was studied. He et al. [22] used CEPT to treat the influent of the urban sewage plant. It was found that the removal effect of CEPT was the best by adding poly aluminum ferric chloride. The COD effluent was always lower than 60mg/L, and the removal rate reached 74 %. In the San Diego sewage treatment plant, the addition of ferric chloride for CEPT treatment achieved a BOD_5 capture rate of 65 % [23]. In addition, CEPT not only has an excellent organic matter capture rate, but also has advantages in organic matter retention. Budych-Gorzna et al. [24] studied the coagulation and sedimentation of primary sludge by using ferric sulfate and carried it out on-site. Experiments showed that the COD removal rate reaches 70 %. In addition, the application of CEPT increases the easily degradable organic matter in sludge digestion, which is beneficial to energy recovery. Rahman et al. [25] found that the COD capture rate of CEPT was 55 %, and the capture rate was not high, but the COD oxidation rate at this stage was 0, and the organic retention reached 100 %.

(3) Other processes

In addition to HRAS and CEPT, membrane separation processes and micro-sieve processes have also begun to be widely used in 'carbon capture' [21]. For example, a highly loaded membrane bio-reactor (MBR) operated at a very short HRT (0.7h) and SRT (0.5 ~ 1d) can recover 85 % of COD and the mineralization rate is less than 10 % [26]. For example, the anaerobic membrane bioreactor (AnMBR) has a COD removal rate of more than 90 % and a CH4 recovery rate of more than 50 % [27]. The micro-screening process uses a fine filtration device (pore 100μm) to intercept particles, colloids, and soluble COD in the sewage. Compared with the blank group, the total chemical oxygen demand (TCOD) removal rate increased by 43 % to 70 %. Its soluble COD accounted for 82 %, an increase of 27 %. Its biogas production reached 150L / kgCOD, an increase of 13 % [21,28].

Compared with the influent quality of municipal sewage in most European countries, the influent COD of most sewage plants in China is lower, which limits the application of front-end carbon capture in sewage plants. However, there is a concept of front-end carbon source separation that applies to our country. Its principle is to recycle cellulose at the front end. Lignocellulose is complex and stubborn, hindering the biodegradation process [29]. Like filamentous bacteria, some 'bridging' effects of lignocellulose may also cause sludge bulking. The screened cellulose has many uses, such as the manufacture of asphalt additives, soil amendments and biocomposites.

3.2 Energy recycling

Energy recovery and utilization is one of the important conditions for achieving carbon neutrality in sewage treatment plants. It is mainly based on energy recovery technologies such as anaerobic digestion-combined heat and power(AD-CHP) and heat recovery.

(1) Anaerobic digestion-combined heat and power(AD-CHP)

Anaerobic digestion technology captures sludge to produce methane (CH_4), which can extract chemical energy from sludge through cogeneration technology. HRAS and CEPT technologies achieve carbon capture. Solon et al. [30] evaluated the CH_4 production of the traditional activated sludge process and HRAS, and found that the CH_4 production of HRAS was 1.5 times that of the traditional activated sludge process. At the same time, Solon et al. also evaluated CEPT and the traditional activated sludge process, and the CH_4 production of CEPT was 25 % higher than that of the traditional activated sludge process. Rahman et al. [25] also found that the methane production of CEPT was 1.8 times that of the traditional activated sludge process. In addition, the energy recovery efficiency depends not only on the amount of carbon source captured, but also on the CH_4 production efficiency of anaerobic digestion. In this regard, researchers have conducted a lot of research to improve the efficiency of anaerobic digestion CH_4 production. Yu [30] used thermal hydrolysis (90 °C) pretreatment to optimize anaerobic digestion, and the average CH_4 yield was 0.159 m^3/kg VS, which was much higher than that of untreated sludge (0.034 m^3/kg VS). The addition of alkali can reduce the temperature of heat treatment and improve the treatment effect. The CH_4 yield was maintained at 0.55 m^3/kg VS by low-temperature hot alkali (60 °C, pH=12.0) pretreatment [31]. Anaerobic co-digestion of other organic solid wastes and excess sludge can also increase CH_4 yield. He [32] co-digested by adding kitchen waste, and the maximum CH_4 yield was 4.3 times higher than that of sludge anaerobic digestion alone. Fitamo et al. [33] conducted a co-digestion pilot test using kitchen waste, and achieved a CH_4 yield of 425 mL/kg VS at medium temperature (55 °C), which was 48 % higher than that of traditional anaerobic digestion. While increasing the CH_4 yield, it also reduces the treatment dilemma of kitchen waste. The biogas produced by anaerobic digestion produces electricity and heat through cogeneration technology. Electric energy can be used for energy self-sufficiency, and thermal energy can be used to supplement digestion tank heating and building heating.

(2) Heat energy recovery

In addition to chemical energy, heat energy is also a recyclable energy in sewage. According to the estimation of the actual recoverable energy in urban sewage, Hao et al. [34] believed that thermal energy is also an energy that cannot be ignored. The sewage temperature is usually below 30 °C, and the sewage flow rate of the sewage treatment plant is stable and available throughout the year, which makes the cold and heat recovery of this low-grade energy practical. At present, heat energy recovery is mainly carried out by the combination of a water source heat pump and heat exchanger. Water source heat pumps and heat exchangers are actually a kind of heating or cooling system. It uses different temperatures between water and the surrounding atmosphere to transfer heat to the water supply or from water, which can effectively extract heat from the water source. Domestic and foreign sewage treatment plants have practical applications in heat energy recovery. For example, the Norwegian Asker sewage treatment plant maintained 15.5 × 104m^2 of commercial building heating by using a water source heat pump [35]. A large municipal sewage treatment plant in Beijing (treatment scale of 1 × 104m^3/d) used a water source heat pump to evaluate the heat recovery. The heat recovery reached 175 GJ at a temperature difference of only 3 °C, which could meet the heating demand of 5 × 104m^2 buildings [36]. A sewage treatment plant in Qingdao not only realizes its own heating demand through a water source heat pump and heat exchanger, but also meets the heating demand of 20 × 104m^2 in the surrounding area [37]. The engineering application shows that the recovered medium-temperature heat energy can be used not only for digestion tank heating, sludge dewatering, and winter heating of the whole plant, but also for heating/cooling of the surrounding areas, which is of great significance for reducing the heating energy consumption of sewage treatment plants and regional heating, and indirectly reducing greenhouse gas emissions. At the same time, it should be noted that the heat energy recovery of raw water will make the heat pump easily damaged, and the treated water is more suitable for heat energy recovery, while avoiding the influence of temperature change on the biochemical system.

(3) Power generation with solar energy

In addition to drawing energy from sewage, solar energy is also available in sewage treatment plants. Large-scale urban sewage treatment plants generally cover a wide area, and various structures have a large surface area. This feature provides the basic conditions for the layout of rooftop photovoltaic power generation technology. A photovoltaic power generation system is arranged above the biochemical tank and the secondary sedimentation tank in a sewage treatment plant, which compensates for 20 % of the total power consumption of the plant [38]. Tangwang and Liuwei sewage treatment plants in Yangzhou City use photovoltaic power generation to achieve energy self-sufficiency [39]. In addition, the coverage of photovoltaic panels on structures may also play a role in thermal insulation, further reducing thermal energy consumption. The combination of anaerobic digestion-cogeneration technology, heat energy recovery, and solar energy can maximize energy recovery and achieve energy self-sufficiency while reducing operating consumption.

The photovoltaic power generation system is arranged above the biochemical tank and the secondary sedimentation tank in the water treatment plant.

3.3 Resource recovery

(1) Phosphorus recovery

Phosphorus is an essential element for maintaining life. It has attracted much attention due to its non-renewable nature. It is difficult to meet the growing uptake of phosphorus by human beings if only relying on the natural cycle of phosphorus. Therefore, it is necessary to develop sustainable phosphorus-use methods to curb the disorderly mining of phosphate rock in nature. Based on the concept of sustainable development, the recovery of phosphorus from sewage has gradually attracted the attention of the industry. The efficient recovery of phosphorus can not only alleviate the shortage of global phosphate rock resources and drive the sustainable development of sewage treatment, but also make up for the deficit of energy consumption in the operation of sewage plants and reduce the emission of greenhouse gases. Although LCA can be used to evaluate phosphorus recovery, most studies have only evaluated economic benefits, and few have evaluated environmental benefits. Compared with traditional fertilizer production, a wastewater treatment plant that uses struvite to achieve full phosphorus recovery can offset about 13 000 t CO_2 emissions per year [40,41]. Chai [42] used LCA to calculate greenhouse gas emissions from sewage treatment plants. The results showed that the recovery of phosphorus and other resources from sewage could effectively reduce greenhouse gas emissions by 7 % ~ 18 %. From the perspective of life cycle, nutrient recovery can not only alleviate the consumption of resources such as phosphate ore, but also indirectly retain energy and water. Recycling nutrients will reduce the demand for traditional fossil fertilizers and save energy and water for the production of traditional fertilizers.

There are two main ways of phosphorus recovery: phosphorus recovery from phosphorus-rich aqueous phase and phosphorus recovery from sludge. The phosphorus-rich aqueous phase recovers phosphorus in the form of magnesium ammonium phosphate (MAP). The phosphate concentration at the end of the anaerobic tank and the sludge digestion solution is high, and the recovery potential is huge [43,44]. Sludge incineration is the main way to recover phosphorus from sludge. The incineration ash contains almost all the phosphorus in the treated sewage, so the recovery is not only simple, but also the maximum recovery (up to 90 % of the phosphorus load of the original sewage) [45]. There are three main methods for phosphorus recovery from sludge ash: biological method, wet chemical method, and thermochemical method. Combining phosphorus-rich aqueous phase and sludge to recover phosphorus can meet the needs of sustainable treatment to the greatest extent. Under the goal of carbon neutralization, this method can not only realize the anaerobic digestion capacity of sludge, but also maximize the recovery of potential resources in sewage. There have been engineering cases of phosphorus recovery from sludge. For example, the Amersfoort wastewater treatment plant in the Netherlands can recover 2 000 t struvite per day from phosphorus-rich concentrate and digestive juice. The purity of the recovered struvite particles is as high as 99.9 %, which can be further packaged as high-quality fertilizer for secondary utilization [8]. Under the background of carbon neutrality, recycling resources and energy from sewage plants have become the trend of industry development, which has brought great development opportunities for the phosphorus recovery industry.

(2) Reclaimed water utilization

After the resources in the sewage are recovered, the reuse of the treated sewage can further improve the utilization rate of water resources and economic benefits. Recycling reclaimed water in the fields of farmland irrigation, industrial production, and landscape water use can reduce the carbon footprint and reduce the energy consumption caused by the development of more energy-intensive water resources. In addition, reclaimed water can also be used for groundwater recharge and direct drinking reuse. Groundwater recharge can alleviate the risk of land subsidence and seawater intrusion in coastal areas. It can also eliminate the demand for ground storage facilities and solve problems related to uncovered surface reservoirs, such as water quality deterioration and odor caused by evaporation loss and algae reproduction. Direct drinking reuse refers to the direct introduction of treated sewage into the water distribution system without involvement in storage (pipeline to pipeline). Reclaimed water reuse can save electricity and chemicals needed to supply water from raw water sources, and has great potential to offset carbon footprint [46]. The main methods of realizing water resource regeneration in sewage plants can be divided into coagulation sedimentation and media filtration. Based on the consideration of the safety of reclaimed water, the reclaimed water utilization technology with membrane treatment as the core has become the development trend of sewage plants. The technology of water resource regeneration and reuse in wastewater treatment plants by membrane treatment has been widely used abroad, which is technically and safely feasible. Singapore plans to achieve the goal of carbon neutrality by 2030 through basic equipment transformation and process transformation of reclaimed water plants. The transformation of basic equipment reduces the energy consumption in the operation process by improving the operational efficiency of the equipment in the reclaimed water plant. There are three main methods for process transformation. One is the side-stream Anammox autotrophic denitrification process, and the other is the membrane bioreactor. The third is sludge pretreatment. The side-stream Anammox autotrophic denitrification process can save the carbon source consumption in the denitrification heterotrophic denitrification process and reduce the aeration energy consumption by 3 % ~ 5 %. The use of a membrane bioreactor (MBR) will increase the power consumption by an additional 0.11 kWh/m^3. However, due to the addition of MBR, the downstream advanced processing units such as microfiltration/ultrafiltration (MF/UF) can be eliminated, and the operational energy consumption will be reduced by 15 %. Even the Zhangyi No.2 reclaimed water plant with double membrane technology (reverse osmosis membrane and microfiltration membrane) has reduced the average power consumption during operation to a low level of 0.75 kWh/m^3 through process optimization [47]. Pretreatment of excess sludge for cell wall breaking can increase the amount of methane produced by anaerobic digestion by 10 % [48]. High-quality reclaimed water has broad market demand in the future and is expected to become an important measure to solve the problem of a water shortage caused by water pollution.

(3) Residual temperature heat energy utilization

For cleaning and washing purposes, the water used in the home is heated to a comfortable temperature or even higher. Due to the input of heat, the average temperature of domestic sewage (27 °C) is 2 ~ 17 °C higher than that of the household [49]. The utilization potential of sewage waste heat energy is higher than the contribution rate of chemical energy in sewage to carbon neutralization (the recovery potential of sewage heat energy is 9 times that of chemical energy) [49]. If 6 % of hot water can be saved or 10 % of heat in sewage can be recovered, the total energy consumed in the sewage treatment process can be compensated [50]. The recovery and utilization of organic chemical energy can be fed back to the sewage treatment plant to make up for the energy deficit. On this basis, the utilization of waste heat energy can achieve the ultimate goal of carbon neutralization through heat output exchange [51]. The waste heat energy can not only meet the needs of the factory, but also radiate to the surrounding houses to meet the needs of refrigeration and heating. The research on the centralized utilization of waste heat energy in Europe was earlier. Since the 1980s, several large-scale water source heat pumps with a total heat capacity of more than 1500 MW have been installed in the heating system in Sweden. These heat pumps are used in regional heating systems [52]. The environmental benefits brought by lower CO_2 emissions provide a broad development prospect for the development of water-source heat pumps. Unlike combustibles such as natural gas, coal, and fuel oil, sewage waste heat energy belongs to low-grade energy and cannot be transported long distances without significant energy loss [53]. Therefore, the heat exchanged by the sewage plant through the water source heat pump is used in situ at the user's end. There are two ways of in-situ utilization, one is home form, and the other is pipeline form. It is worth noting that the in-situ utilization will cause serious unit corrosion during operation. Therefore, the heat exchanger should be inspected and repaired regularly to solve the problems of pollution, blockage, and corrosion in time. The heat output of the water source heat pump needs to be consumed by users with large and stable heat demand, so the exchange of heat for sludge thermal drying is a potential choice. The water content of sludge can be reduced from 80 % to 40 % ~ 70 % by using sewage waste heat energy for sludge drying. From the perspective of sustainable development of sewage treatment plants, sludge incineration after drying realizes the conversion of waste heat energy from low-grade energy to high-grade energy. It is of great practical significance to recover heat energy from sewage. Reducing the carbon footprint will be the only way for the development of sewage treatment plants in the future. Through the recovery and utilization of waste heat energy in sewage, more carbon emission reductions will undoubtedly be achieved, to achieve the goal of carbon neutralization operation.

4. Conclusions

The key to achieving the goal of carbon neutrality in wastewater treatment plants is to recycle resources and energy and save energy and reduce the consumption of treatment facilities. At present, China's sewage plants still face some challenges in achieving carbon neutrality, which are mainly divided into the following three aspects: (1) The influent COD of sewage treatment plants in China is generally low, which is seriously limited. (2) In China, less than 3 % of wastewater treatment plants are equipped with anaerobic digestion facilities, and a considerable part of them are in poor operation. (3) Most of the sewage treatment plants in China have a low degree of intelligence and fine control, and it is impossible to adjust the dosage and aeration amount in real-time according to the influent load.

Therefore, the future development direction is mainly the following four aspects: (1) Strengthen the development of low-carbon and resource recoveries and treatment technologies, such as anaerobic ammonia oxidation technology, short-cut nitrification, and denitrification technology, to minimize the overall energy balance of the sewage plant. (2) Improve the level of pipe network construction, improve sewage collection rate and influent COD, strengthen sludge resource disposal, and introduce an exogenous organic matter for anaerobic co-substrate digestion to improve energy recovery efficiency. For example, the use of food waste for anaerobic co-digestion can increase the organic matter content in the digestion process and increase the biogas yield. (3) Make full use of the waste heat energy in sewage, develop efficient heat exchange technology, and promote the development of recycling heat energy. (4) Accelerate the intelligent development and automation process of sewage treatment plants, and realize the transformation of the operation mode of urban sewage treatment plants.

Foreign sewage treatment plants have achieved carbon-neutral operation, which provides a reference for the operation of sewage treatment plants in China. To achieve carbon neutrality, wastewater treatment plants need to efficiently recycle resources and energy to achieve energy self-sufficiency. Resource-based sewage treatment is gradually becoming the theme of the times in the global sewage treatment industry. Although China's sewage plants still face a series of challenges in achieving carbon neutrality, they are also actively exploring ways suitable for their development. The concept of sewage treatment plants has opened a new chapter of sustainable management of sewage and harmonious urban water ecology in China.

References

1. DAI Xiaohu, ZHANG Chen, ZHANG Linwei, et al. Thoughts on the development direction of sludge treatment and resource recovery under the background of carbon neutrality[J]. Water & Wastewater Engineering, 2021, 57(3): 1-5.
2. SHAN Yuli, HUANG Qi, GUAN Dabo, et al. China CO2 emission accounts 2016—2017[J]. Scientific Data, 2020, 7(1): 54.
3. NIU Kunyu, WU Jian, QI Lu, et al. Energy intensity of wastewater treatment plants and influencing factors in China[J]. Science of the Total Environment, 2019, 670: 961-970.
4. LOOSDRECHT M, BRDJANOVIC D. Anticipating the next century of wastewater treatment[J]. Science, 2014, 344(6191): 1452-1453.
5. GONG Hui, JIN Zhengyu, XU Heng, et al. Redesigning C and N mass flows for energy-neutral wastewater treatment by coagulation adsorption enhanced membrane (CAEM)-based pre-concentration process[J]. Chemical Engineering Journal, 2018, 342: 304-309.
6. HAO Xiaodi, WEI Jing, CAO Yali.A successful case of carbon-neutral operation in America: Sheboygan WWTP s[J]. China Water&Wastewater, 2014, 30(24): 1-6.
7. HAO Xiaodi, JIN Ming, HU Yuansheng. Framework of future waste-water treatment in the Netherlands: NEWs and their practices[J]. China Water&Wastewater, 2014, 30(20): 7-15.
8. HAO Xiaodi, REN Bingqian, CAO Yali. An engineering model of sustainable wastewater treatment: Steinhof WWTP at Braunsch-weig in Germany[J]. China Water&Wastewater, 2014, 30(22): 6-11.
9. NOWAK O, ENDERLE P, VARBANOV P. Ways to optimize the energy balance of municipal wastewater systems: Lessons learned from Austrian applications[J]. Journal of Cleaner Production, 2015, 88: 125-131.
10. QU Jiuhui, WANG Hongchen, WANG Kaijun, et al. Municipal wastewater treatment in China: Development history and future perspectives[J]. Frontiers of Environmental Science&Engineering, 2019, 13(6): 1-7.
11. GUO Chaoran, HUANG Yong, ZHU Wenjuan, et al. Organics recovery from municipal wastewater: research advances in capture technologies[J]. Chemical Industry and Engineering Progress, 2021, 40(3): 1619-16335.
12. RAHMAN A, MEERBURG F A, RAVADAGUNDHI S, et al. Bioflocculation management through high-rate contact-stabilization: a promising technology to recover organic carbon

from low-strength wastewater[J]. Water Research, 2016, 104: 485-494.

13. ALVARADO V I, HSU S C, LAM C M, et al. Beyond energy balance: environmental trade-offs of organics capture and low carbon-to-nitrogen ratio sewage treatment systems[J]. Environmental Science&Technology, 2020, 54(8): 4746-4757.
14. SARPONG G, GUDE V G. Near future energy self-sufficient wastewater treatment schemes[J]. International Journal of Environmental Research, 2020, 14(4): 479-488.
15. HUANG Baocheng. Recovery and versatile reuse of organic carbon from municipal wastewater[D]. Hefei: University of Science and Technology of China, 2018.
16. LU Xinxin. Fundamental research of low consumption domestic sewage treatment technology based on enhanced primary treatment[D]. Xi'an: Xi'an University of Architecture and Technology, 2020.
17. SANCHO I, LOPEZ-PALAU S, ARESPACOCHAGA N, et al. New concepts on carbon redirection in wastewater treatment plants: a review [J]. Science of the Total Environment, 2019, 647: 1373-1384.
18. CHEN Jialiang. Experimental research on separating organic carbon from municipal wastewater by high-rate activated sludge system[D]. Tangshan:North China University of Science and Technology, 2020.
19. [GUVEN H, ERSAHIN M E, DERELI R K ,et al. Effect of hydraulic retention time on the performance of high-rate activated sludge system: a pilot-scale study[J]. Water, Air&Soil Pollution, 2017, 228(11): 1-10.
20. GUVEN H, FAKIOGLU M, SINOP I, et al. Retrofitting of five preliminary wastewater treatment plants in Istanbul (Turkey)to high-rate activated sludge system and/or post oxidation[J]. Ozone: Science&Engineering, 2020, 42(3): 255-266.
21. LIU Zhixiao. Carbon capture and carbon redirection:new way to optimize the energy self-sufficient of wastewater treatment[J]. China Water&Wastewater, 2017, 33(8): 43-52.
22. HE Qiulai, WANG Hongyu, XU Congyuan, et al. Feasibility and optimization of wastewater treatment by chemically enhanced primary treatment(CEPT): a case study of Huangshi[J]. Chemical Speciation& Bioavailability, 2016, 28(1/2/3/4): 209-215.
23. NUNEZ C, DORNFELD M, SHANKLES K C, et al. Cost savings and performance improvement of large system iron salt use for integrated sulfide control and chemically enhanced primary treatment by using peroxide regenerated iron technology[J].Proceedings of the Water Environment Federation, 2010, 2010(16): 1110-1121.
24. BUDYCH-GORZNA M, SZATKOWSKA B, JAROSZYNSKI L, et al. Towards an energy self-sufficient resource recovery facility by improving energy and economic balance of a municipal WWTP with chemically enhanced primary treatment[J]. Energies, 2021, 14(5): 1445.
25. RAHMAN A, DE CLIPPELEIR H, THOMAS W, et al. A-stage and high-rate contact-stabilization performance comparison for carbon and nutrient redirection from high-strength municipal wastewater[J]. Chemical Engineering Journal, 2019, 357: 737-749.
26. FAUST L, TEMMINK H, ZWIJNENBURG A, et al. High loaded MBRs for organic matter recovery from sewage:effect of solids retention time on bioflocculation and on the role of extracellular polymers[J]. Water Research, 2014, 56: 258-266.
27. WAN Junfeng, GU Jun, ZHAO Qian, et al. COD capture:a feasible option towards energy self-sufficient domestic wastewater treatment[J]. Scientific Reports, 2016, 6: 25054.
28. SOLON K, JIA Mingsheng, VOLCKE E I P. Process schemes for future energy-positive water resource recovery facilities[J]. Water Science and Technology: A Journal of the International Association on Water Pollution Research, 2019, 79(9): 1808-1820.
29. HENDRIKS A, ZEEMAN G. Pretreatments to enhance the digest-ibility of lignocellulosic biomass[J]. Bioresource Technology, 2009, 100(1):10-18.
30. YU Chuandai. Methane production from semi-continuous anaerobic digestion reactor using municipal sludge pretreated by low temperature thermal hydrolization[D]. Fuzhou: Fujian Normal University, 2017.
31. XIAO Benyi, TANG Xinyi, YI Hao, et al. Comparison of two advanced anaerobic digestions of sewage sludge with high-temperature thermal pretreatment and low-temperature thermal-alkaline pretreatment[J]. Bioresource Technology, 2020, 304: 122979.
32. HE Meilong. Methanogenesis performance enhancement of municipal sludge anaerobic digestion based on substrate conditioning[D]. Fuzhou: Fujian Normal University, 2018.
33. FITAMO T, BOLDRIN A, BOE K, et al. Co-digestion of food and garden waste with mixed sludge from wastewater treatment in continuously stirred tank reactors[J]. Bioresource Technology, 2016, 206: 245-254.
34. HAO Xiaodi, LI Ji, VAN LOOSDRECHT M C M, et al. Energy recovery from wastewater:heat over organics[J].Water Research, 2019, 161: 74-77.
35. SONG Xinxin, LIN Jia, LIU Jie, et al. The current situation and engineering practice of sewage treatment technology facing the future[J]. Acta Scientiae Circumstantiae, 2021 ,41(1): 39-53.
36. HAO Xiaodi, LIU Ranbin, HUANG Xin. Evaluation of the potential for operating carbon neutral WWTPs in China[J]. Water Research, 2015, 87: 424-431.
37. LIU Ruling, SONG Peng, DAI Weidong. Application of sewage-source heat pump technology in Qingdao

Tuandao sewage treatment plant[J]. China Water&Wastewater, 2015, 31(12): 86-89.

38. ZHANG Kaihai. Application of distributed photovoltaic power generation system in a wastewater treatment plant[J]. China Water&Wastewater, 2017, 33(22): 81-84.
39. LIU Xiaoming, YAN Junquan, HUANG Penglan. Innovative application of solar power generation in water treatment industry[J]. China Water&Wastewater, 2015, 31(18): 90-94.
40. MO Weiwei, ZHANG Qiong. Energy-nutrients-water nexus: Integrated resource recovery in municipal wastewater treatment plants[J]. Journal of Environmental Management, 2013, 127: 255-267.
41. NAKAKUBO T, TOKAI A, OHNO K. Comparative assessment of technological systems for recycling sludge and food waste aimed at greenhouse gas emissions reduction and phosphorus recovery[J]. Journal of Cleaner Production, 2012, 32: 157-172.
42. CHAI Chunyan. Study on the characteristics of greenhouse gas emissions and heat island effect of municipal wastewater treatment plants[D]. Harbin: Harbin Institute of Technology, 2017.
43. CORNEL P, SCHAUM C. Phosphorus recovery from wastewater: Needs, technologies and costs[J]. Water Science&Technology, 2009, 59(6): 069-1076.
44. ZHOU Kaixin, BARJENBRUCH M, KABBE C, et al. Phosphorus recovery from municipal and fertilizer wastewater: China's potential and perspective[J]. Journal of Environmental Sciences, 2017, 52(2): 151-159.
45. [HAO Xiaodi, YU Jinglun, LIU Ranbin, et al. Advances of phosphorus recovery from the incineration ashes of excess sludge and its associated technologies[J]. Acta Scientiae Circumstantiae, 2020, 40(4): 1149-1159.
46. MENESES M, PASQUALINO J C, CASTELLS F. Environmental assessment of urban wastewater reuse: Treatment alternatives and applications[J]. Chemosphere, 2010, 81(2): 266-272.
47. LU Ruiqing, YANG Guang, GONG Hui, et al. Enlightenment of Singapore's NEWater technology to the production of high quality reclaimed water in China[J]. China Water&Wastewater, 2019, 35(14): 36-40.
48. HAO Xiaodi, MENG Xiangting, FU Kunming. Targeted energy consumption and associated technologies developed in water reclamation plants in Singapore[J]. China Water&Wastewater, 2014, 30(24): 7-11.
49. HOFMAN J, HOFMAN-CARIS R, NEDERLOF M, et al. Water and energy as inseparable twins for sustainable solutions[J]. Water Science and Technology, 2011, 63(1): 88-92.
50. HAO Xiaodi, LI Ji, VAN LOOSDRECHT M C M, et al. Energy recovery from wastewater: Heat over organics[J]. Water Research, 2019, 161: 74-77.
51. HAO Xiaodi, LIU Raibin, HUANG Xin. Evaluation of the potential for operating carbon neutral WWTPs in China[J]. Water Research, 2015, 87: 424-431.
52. AVERFALK H, INGVARSSON P, PERSSON U, et al. Large heat pumps in Swedish district heating systems[J]. Renewable&Sustainable Energy Reviews, 2017, 79: 1275-1284.
53. SPRIET J, MCNABOLA A, NEUGEBAUER G, et al. Spatial and temporal considerations in the performance of wastewater heat recovery systems[J]. Journal of Cleaner Production, 2020, 247: 119583.

Research on the driving characteristics of motor vehicles in high-altitude areas for the analysis of exhaust emission characteristics

Zhaoyu Qi [1], Ming Gu [1], Chuanzhou You [2], Yue Zhan [2], Zhongwu Ma [2], Wei Huang [1,*]

[1] Key Laboratory of Environmental Protection in Water Transport Engineering Ministry of Transport, Tianjin Research Institute for Water Transport Engineering, M.O.T., Tianjin, 300456, China
[2] Qinghai Transportation Planning & Design Institute Co. Ltd., Xining, 810001, China

Abstract: Research on the driving characteristics of motor vehicles at different altitudes in plateau regions and road emission intensity can provide technical support for energy conservation and emission reduction policies in plateau areas. This paper selects gasoline and diesel vehicles for on-board testing on typical road sections in plateau regions to study the overall driving characteristics of motor vehicles and the driving characteristics at different altitude intervals in plateau areas. The results show that with the increase in altitude, the driving conditions gradually concentrate towards 'high speed, low acceleration.' The relevant conclusions can further study and analyze the distribution patterns of pollutants in motor vehicle exhaust emissions and road emission intensity.

Keywords: high-altitude; motor vehicle; driving characteristics; energy conservation and emission reduction; on-board testing

1. Introduction

Plateau regions exhibit significant differences in environmental factors such as atmospheric pressure, oxygen concentration, and temperature compared to plain areas due to variations in altitude [1]. Therefore, studying the driving characteristics of motor vehicles and road emission intensity at different altitudes in plateau regions can provide support for government departments in formulating energy conservation and emission reduction policies specifically for plateau areas. This includes the development and adjustment of policies such as vehicle restrictions, traffic control, and prioritization of public transportation in plateau regions [2]. Automobile manufacturers can optimize vehicle design for the unique environmental conditions of plateau regions based on the driving characteristics and emission intensity at different altitudes, aiming to improve combustion efficiency, reduce carbon emissions, and enhance the market competitiveness of automobiles in plateau areas [3].

In this study, two vehicle types with high ownership and activity levels in plateau regions were selected as typical vehicles for on-board testing, namely diesel-powered vehicles compliant with the National IV emission standard and gasoline-powered vehicles compliant with the National V emission standard. The test route includes major road types in plateau regions, reflecting real-world driving patterns of vehicles, as well as variations in altitude and terrain. The starting point of the test route is Chengbei District, Xining City, Qinghai Province (altitude: 2270m), and the endpoint is the toll station in Maduo County, Golog Tibetan Autonomous Prefecture (altitude: 4540m), with a total distance of 472 kilometers and an altitude span of approximately 2400 meters.

2. Parameter selection for analysis

The driving behavior of motor vehicles is typically described using various statistical parameters to characterize their overall motion characteristics, such as travel time, driving distance, average speed, maximum acceleration, etc. In this study, the average speed (V) and acceleration (a) are chosen as the statistical feature parameters to describe the driving behavior of motor vehicles. Their calculation formulas are as follows:

* Corresponding author: huangwei@tiwte.ac.cn

$$v = \frac{D}{T}$$

where D represents the total driving distance in meters (m), and T represents the total travel time in seconds (s).

$$a_i = \frac{v_{i+\Delta t} - v_{i-\Delta t}}{2\Delta t}$$

where a_i represents the acceleration of motor vehicle at time second i, m/s^2; $\triangle t$ represents the time interval between i and $i+\triangle t$, $\triangle t$ is set to 1 second (s).

3. Overall driving characteristics

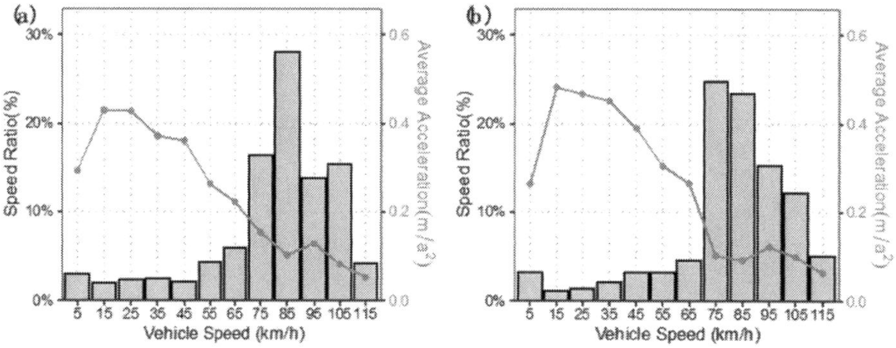

Figure 1 The driving characteristics of a diesel vehicle (a) and a gasoline vehicle (b)

Figure 1 represents the distribution of speed and acceleration during two testing sessions. As shown in Figure 1, the driving speeds of both vehicles are primarily distributed above 75 km/h, accounting for approximately 80% of the total travel time. This is because the driving routes mainly consist of highways where vehicles can attain higher speeds. The routes also pass through national nature reserves and frozen soil roads with strict speed limits, ensuring that the maximum driving speed does not exceed 120 km/h.

Regarding acceleration, the average acceleration is higher during low-speed driving intervals and lower during high-speed driving intervals. This is due to the relatively low traffic volume in the test route, which prevents traffic congestion. During low-speed driving, frequent braking and acceleration occur due to road bumps or wildlife avoidance, resulting in a higher average acceleration.

4. Driving characteristics at different altitude intervals

To analyze the variations in driving behavior at different altitude regions, the research route was divided into four altitude intervals based on 500m increments: [2000, 2500m), [2500, 3000m), [3000, 3500m), [3500, 4000m). Figures 2 and 3 depict the distribution of speed and acceleration for diesel and gasoline vehicles in different altitude intervals. The average speeds of the diesel vehicle in the four altitude intervals are 17.2 km/h, 22.8 km/h, 23.8 km/h, and 21.5 km/h, with average absolute accelerations of 0.2 m/s², 0.13 m/s², 0.12 m/s², and 0.16 m/s², respectively. The average speeds of the gasoline vehicle in the four altitude intervals are 16.2 km/h, 24.9 km/h, 23.2 km/h, and 21.5 km/h, with average absolute accelerations of 0.23 m/s², 0.11 m/s², 0.16 m/s², and 0.10 m/s², respectively.

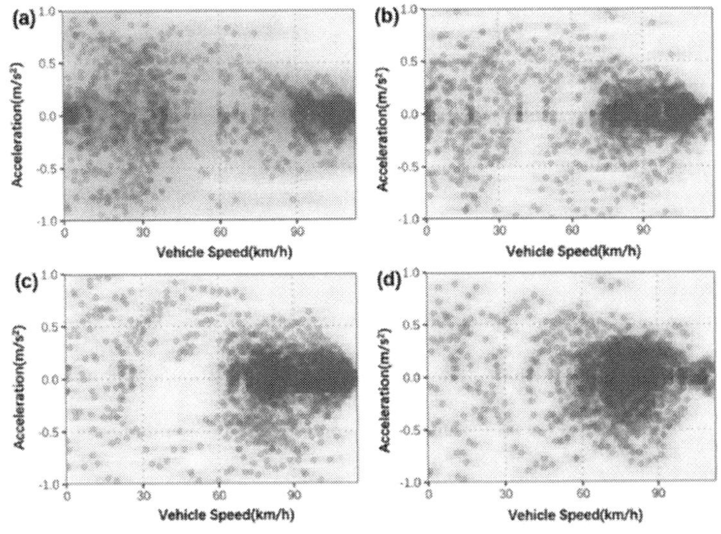

Figure 2 The speed-acceleration distribution of the diesel vehicle at different altitudes:

(a)2000-2500m(b)2500-3000m(c)3000-3500m(d)3500-4000m

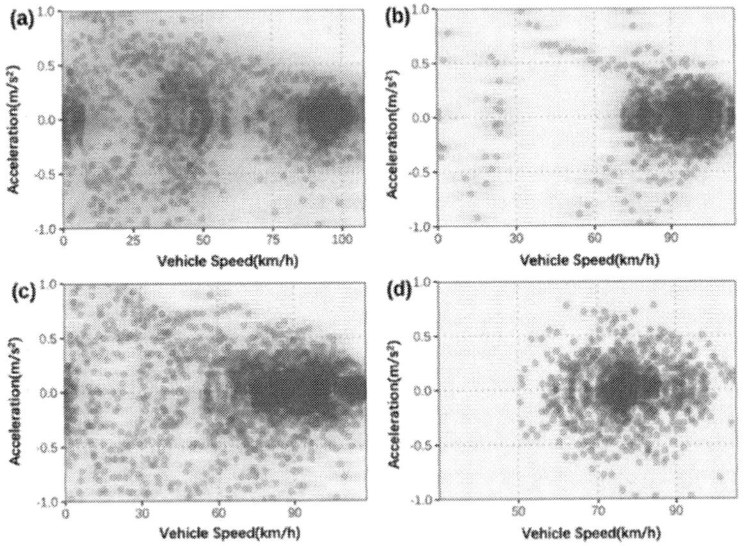

Figure 3 The speed-acceleration distribution of the gasoline vehicle at different altitudes:

(a)2000-2500m(b)2500-3000m(c)3000-3500m(d)3500-4000m

From Figures 2 and 3, it can be observed that as the altitude increases, the driving conditions gradually concentrate towards "higher speeds and lower accelerations." This is because as the altitude rises, the flow of traffic and pedestrians decreases, resulting in smoother traffic conditions compared to lower altitude areas, which in turn reduces the frequency of acceleration and deceleration, leading to more stable driving conditions. The altitude interval of 2000-2500m (Figures 2a and 3a) mainly includes roads in Xining city and national/provincial highways, where vehicles are in the starting phase and frequently encounter intersections and congested sections, resulting in a relatively scattered distribution of driving characteristics in this altitude range. In terms of typical motor vehicle exhaust emissions, HC emissions are mainly caused by incomplete combustion due to rapid acceleration and the occurrence of engine stall during high-speed deceleration. CO emissions primarily result from incomplete combustion during acceleration. NO_x emissions mainly stem from excessively high combustion temperatures during acceleration [4]. The road emission intensity of pollutants from motor vehicle exhaust shows significant variations depending on road conditions and the types of gases emitted [5]. Therefore, by combining the relevant conclusions on the driving characteristics of motor vehicles in high-altitude regions presented in this study, further research can be conducted to analyze the distribution patterns of pollutants in motor vehicle exhaust emissions and road emission intensity.

5. Conclusions

This study investigated the overall driving characteristics and driving patterns at different altitude ranges for motor vehicles in plateau regions, including parameters such as speed, acceleration, and average absolute acceleration. The results indicate that in high-altitude areas, where there is less traffic and fewer pedestrians, vehicles tend to travel at higher speeds and maintain stable driving conditions. As the altitude increases, the driving conditions gradually shift towards a concentration of "higher speeds and lower accelerations." By combining the findings on the driving characteristics of motor vehicles in high-altitude regions presented in this study, further research can be conducted to analyze the distribution patterns of pollutants in motor vehicle exhaust emissions and road emission intensity.

Acknowledgments

The authors gratefully acknowledge financial support by the Qinghai Provincial Key R&D and Transformation Plan (Grant No. 2022-SF-149), Qinghai Transportation Science and Technology Project (Grant No. 2023-05), and the Basic Research Fund of Central Public-interest Scientific Research Institutes in China (Grant No. TKS20230301).

References

1. Wang, Y., Feng, X., Zhao H., et.al. (2022). Experimental study of CO2 and pollutant emission at various altitudes: Inconsistent results and reason analysis. Fuel, 307: 121801.
2. Giraldo, M., Huertas J I. (2019). Real emissions, driving patterns and fuel consumption of in-use diesel buses operating at high altitude. Transportation Research Part D: Transport and Environment, 77: 21-36.
3. Liu, J., Li, Y., Zhang, C., et.al. (2022). The effect of high altitude environment on diesel engine performance: Comparison of engine operations in Hangzhou, Kunming and Lhasa cities. Chemosphere, 309: 136621.
4. Agarwal A K., Mustafi N N.. (2021). Real-world automotive emissions: Monitoring methodologies, and control measures. Renewable and Sustainable Energy Reviews, 137: 110624.
5. Xu, Y., Liu, Z., Xue, W., et.al. (2021). Identification of on-road vehicle CO2 emission pattern in China: A study based on a high-resolution emission inventory. Resources Conservation and Recycling, 175: 105891.

Understanding of the Tracking and Adjusting Effect of Deep Profile Control and Flooding Expansion Application Blocks in Pubei Oilfield

Guoguang Wang

The Seventh Oil Production Plant of Daqing Oilfield Co., Ltd. in Heilongjiang Province, Daqing, China

Abstract: In order to explore the development technology of enhancing oil recovery during the ultra-high water cut period in the Pubei Oilfield, deep profile control pilot tests and expanded application tests have been carried out in the Pubei Oilfield. This article mainly starts from the actual production performance of the deep profile control and flooding expansion application test, analyzes the main tracking and adjustment methods and the understanding of the results obtained in the development process of deep profile control and flooding, laying the foundation for the establishment of deep profile control and flooding matching tracking and adjustment technology in Pubei Oilfield.

Keywords: Deep profile control and flooding; Track and adjust; Effect recognition.

1. Introduction

The on-site testing of oil and water wells in the expanded application block of deep profile control and flooding in Pubei Oilfield was successively put into operation and injected in January 2018, and is currently in the late stage of main slug injection. In order to fully leverage the effectiveness of deep profile control and flooding oil production development, and explore the matching methods for tracking and adjusting deep profile control and flooding in Pubei Oilfield, a full process tracking and adjustment method has been developed during the implementation of deep profile control and flooding plans to ensure the effectiveness of profile control and flooding development; Classify and optimize injection plans to maintain a stable increase in injection pressure; Develop parameter adjustment technical boundaries to ensure smooth injection; Classify and optimize measures to reduce pressure and increase injection capacity; Timely adjustment work in oil well fracturing transformation, improving the development effect of profile control and flooding, etc. By summarizing the results and understanding, a set of deep profile control and flooding matching adjustment technologies suitable for the northern region of Portugal is formed to guide the production of the block and provide technical support for the stable production of the oilfield.

2. Overview of Block

The experimental area for expanding the application of deep profile control and flooding is located in the middle of the third fault block in the Pubei Oilfield. The development layer is the PuI Formation oil layer, divided into 11 small layers and 26 sedimentary units. The experimental area has an oil bearing area of 1.45km^2 and a geological reserve of 132.77 × 10^{4}t, average air permeability 298 × 10^{-3} μm^2, pore volume 271.48 × 10^{4}m^3, with a total of 43 wells, including 15 injection wells and 28 production wells. The average single well sandstone thickness is 12.9m, and the effective thickness is 7.9m. The scheme selects composite ion polymer gel system for injection, and the designed injection pore volume is 0.30PV. The test entered the blank water drive stage in January 2018. In September 2018, the pre slug injection of deep profile control and drive system was implemented. In January 2019, the main slug injection of deep profile control and drive system was started. By June 2023, 69.33 profile control and drive agent had been accumulatively injected × 10^{4}m^3, injected into underground pore volume of 0.255PV, with a cumulative oil increase of 5.67 in stages × 10^{4}t, increased oil recovery by 4.28 percentage points.

3. Main methods for tracking and adjusting

3.1 Strengthen dynamic tracking and adjustment to improve the effect of displacement control

3.1.1 Develop a full process tracking and adjustment method to ensure the effectiveness of profile control and flooding development

Drawing on the experience gained from the Pu125 pilot test of deep profile control and flooding in Pubei Oilfield, this study explores the tracking and adjustment methods for the entire process of deep profile control and flooding. Based on different water bearing stages of profile control and flooding, adjustment goals and methods are formulated and optimized to ensure the development effect of deep profile control and flooding. Regarding the ineffective period: due to the rapid increase in injection pressure, the oil well is not significantly affected. Therefore, high concentration and low speed injection should be maintained appropriately to achieve the goal of increasing injection pressure and expanding the affected volume; For the period of water cut decrease: the injection pressure steadily increases, and the oil well begins to take effect. Timely adjustment of well group parameters, timely stimulation of oil well fracturing, and well measures to reduce pressure and increase injection are carried out to achieve the goal of balanced effect and water cut reduction; For the low water content period: individual wells have increased injection pressure and maintained low water content, optimizing injection and production parameters, increasing oil well fracturing efficiency, and reducing water well pressure measures to achieve the goal of maintaining injection and production balance and extending the low water content period; In response to the recovery period of water content: the overall injection pressure is high, and the water content slowly rises. Adhere to the combination of improvement and control, plane adjustment, and well group potential tapping and governance to achieve the goal of maintaining effective injection and controlling the recovery of water content; For subsequent water flooding: the injection pressure is basically stable, and the water content gradually rises. Stratified water injection is carried out in an orderly manner, and the injection production structure adjustment is optimized to achieve the goal of controlling interlayer and plane contradictions.

3.1.2 Classify and optimize injection plans to maintain a stable increase in injection pressure

Personalize injection plans based on reservoir development characteristics and production dynamics, guide optimization and adjustment of injection parameters, and ensure effective injection of the system. In the early and middle stages of injection, injection wells are divided into two categories: Class I wells with large reservoir development thickness, relatively high permeability and connectivity ratio, and large injection pressure rise space adopt the adjustment method of "maintaining concentration and stable injection"; Wells with relatively small reservoir development thickness, low permeability and connectivity ratio, and high injection pressure in Class II wells adopt the adjustment method of "low concentration and low speed injection".

3.1.3 Develop parameter adjustment technical boundaries to ensure smooth injection

Based on the experience of mathematical model prediction and parameter adjustment for different water bearing stages, the control principles and boundaries of injection rate, concentration, and flow pressure parameters are formulated to guide the optimization and adjustment of injection production parameters, ensuring smooth injection.

During the non effective period and water cut decreasing period, the adjustment principle of "low speed and high concentration injection, high water cut and high flow pressure, and low water cut and low flow pressure" is adopted to promote balanced efficiency of the reservoir and maintain a stable increase in pressure.

During the low water content period, the adjustment principle of "maintaining concentration and velocity injection, and producing at low flow pressure" is adopted to ensure balanced injection and production, and improve the effect of profile control and flooding.

3.2 Carry out measures to finely tap potential and expand the development effect of profile control and flooding

3.2.1 Classification optimization of depressurization and injection increase measures to improve injection capacity

In the process of deep profile control and flooding, wells with large fluctuations in injection volume should be washed with high displacement to timely remove blockages in the wellbore and near wellbore areas, and maintain stable injection; For wells with poor reservoir development, poor depressurization effect of conventional measures, allowable injection pressure difference less than 0.5MPa, high initial injection pressure, and poor later well flushing effect, large-scale fracturing methods such as deep penetration fracturing and multi fracture fracturing are adopted to improve reservoir conditions and improve injection capacity. A total of 163 well renovations were implemented, including 153 well washes and 10 well fracturing.

3.2.2 Timely oil well fracturing transformation to improve the development effect of profile control and flooding

In the process of deep profile control and flooding, it is recommended to select effective wells or trend wells with a water content reduction of more than 3%, a liquid production reduction of more than 20%, a profile control and flooding thickness of more than 4.0m, and an effective thickness greater than 1.0m in the main reservoir. Timely oil well fracturing and efficiency enhancement are carried out to further improve the effectiveness of deep profile control and flooding.

A total of 13 oil well fracturing operations were carried out, with an average daily increase of 37.7 tons of fluid and 5.2 tons of oil per day in the initial stage. The water content decreased by 3.1 percentage points. As of now, the average single well measure has increased oil by 1939 tons, with an effective period of 532 days. Based on the actual results of fracturing and summarizing experience and understanding, the requirements for deep profile control and flooding oil well fracturing technology have been formed.

4. Achieve stage test results and understanding

4.1 Maintain good injection and production capacity

In the initial stage of injection, adhere to low-speed injection, and maintain the apparent suction index at around $0.44m^3/(d \cdot m \cdot MPa)$, a decrease of 21.4% in injection capacity compared to before profile control. In the middle and later stages of injection, through optimization of injection parameters adjustment and fracturing modification, the injection capacity remains basically stable, with a slight decrease.

Since the deep profile control and flooding experiment, the daily liquid production has been relatively low in the initial stage. In the middle and later stages, supporting treatment measures such as optimizing parameter adjustments and fracturing measures have gradually improved the liquid production capacity of the experimental area. Currently, the daily liquid production capacity is maintained at around 750t, the liquid production index is maintained at around $0.80t/(d \cdot m \cdot MPa)$, and the liquid production capacity is basically stable.

4.2 Improvement of oil reservoir utilization degree

The utilization of oil layers has been significantly improved after deep profile control and flooding. According to the analysis of continuous injection profile data from 15 injection wells, the effective utilization ratio of the oil layer in the initial injection stage has increased to 72.8%, which is 9.7 percentage points higher than before deep profile control and flooding. Currently, the effective utilization ratio of the oil layer is stable at around 80.2%, and reservoirs with different permeability levels have been utilized. Among them, $150 \times 10^{-3}\mu m^2$ The utilization degree of medium and high permeability reservoirs above is significantly higher than $150 \times 10^{-3}\mu m^2$ Low permeability reservoirs below, and high permeability reservoirs have reduced relative liquid absorption, transferring to medium and low permeability reservoirs, and the proportion of utilization in medium and low permeability reservoirs has increased significantly.

4.3 Significant effect of increasing oil and reducing precipitation

After the measures were taken, the daily oil production of the entire region increased by 24.0 tons, and the water content decreased by 1.9 percentage points, with a cumulative increase of 5.67% during the stage × 104t, increased oil recovery by 4.28 percentage points, and had a significant effect on increasing oil and precipitation. The proportion of effective wells is high, with a total of 27 produced wells in the entire area showing significant results, with an effective rate of 96.4%. Compared with before the effectiveness, the average daily oil production of a single well in the entire area increased by 0.5t, and the water content decreased by 2.0 percentage points.

4.4 Phase effect better than expected in the plan

Currently, the cumulative oil increase in the experimental area is 5.67 × 104t, with a stage increase of 4.28 percentage points in oil recovery rate. Overall, the stage effect is better than the plan's expectation, with a recovery degree of 0.05 percentage points higher than the plan's expectation, and an additional 0.07 percentage points of oil increase × 104t, with a water content lower than the expected 0.57 percentage points in the plan.

5. Conclusion

(1) Dynamic tracking adjustment is an effective way to improve the effectiveness of deep profile control and flooding. Timely carry out the development of full process tracking and adjustment methods, classify and optimize injection plans, develop parameter adjustment technical boundaries, establish dynamic tracking and adjustment charts, and other adjustment work, in order to seize favorable opportunities and maximize the role of deep water flooding.

(2) Measures to tap potential are a powerful guarantee for expanding the effectiveness of profile control and flooding development. In the process of deep profile control and flooding, high displacement flushing and fracturing of injection wells are effective means to improve injection capacity; Timely implementation of fracturing transformation on the produced wells can significantly improve the effectiveness of deep profile control and flooding.

(3) Deep profile control and flooding is an effective means to tap the potential of remaining oil during the ultra-high water cut development period. Since the deep profile control and flooding, the injection and production capacity of the experimental area has remained good, the

degree of oil reservoir utilization has increased, and the effect of increasing oil and precipitation is obvious. The stage increase in oil recovery rate is 4.28 percentage points, and the stage effect is better than the expected plan.

References

1. Yuan Shuwu, Han Jiasheng, Saruli Caokti, Yuan Shuai, Sha Rong. Application of deep profile control and flooding technology in secondary development of conglomerate reservoirs. Xinjiang Petroleum and Natural Gas. June 2015, Volume 11, Issue 2
2. Xu Jianjun, Huang Lida, Yan Limei, Yi Na. Insulator Self-Explosion Defect Detection Based on Hierarchical Multi-Task Deep Learning[J]. Transactions of China Electrotechnical Society, 2021,36(07):1407-1415.
3. Yan Limei, Liu Yongqiang, Xu Jianjun, et al. Broken string diagnosis of composite insulator based on Grabcut segmentation and filler area discrimination[J]. Power System Protection and Control, 2021, 49(22): 114-119
4. Yi, Q. Wang, L. Yan, et al., A multi-stage game model for the false data injection attack from attacker' s perspective. Sustainable Energy Grids & Networks 28 (2021).
5. Na Yi, Jianjun Xu, Limei Yan, Lin Huang. Task Optimization and Scheduling of Distributed Cyber-physical System Based on Improved Ant Colony Algorithm. Future Generation Computer Systems, 109(Aug. 2020), 134-148.
6. Yang Zhao, Jianjun Xu, Jingchun Wu. A New Method for Bad Data Identification of Oilfield Power System Based on Enhanced Gravitational Search-Fuzzy C-Means Algorithm. IEEE Transactions on Industrial Informatics. VOL. 15, NO. 11, NOVEMBER 2019 5963-5970
7. Jing Han, Xi Wang, LiMei Yan, Aida Dahlak, et al. Modelling the performance of an SOEC by optimization of neural network with MPSO algorithm. International Journal of Hydrogen Energy, Volume 44, Issue 51, 22 October 2019, Pages 27947-27957.https://doi.org/10.1016/j.ijhydene.2019.09.055

Research on the Development Change Law of Block A after Fine Potential Excavation

Pan Guo

No.5 Production Plant of Daqing Oil Field, Heilongjiang Province, Daqing, China

Abstract. After the comprehensive utilization of well pattern and fine potential tapping based on subdivision water injection in block A, remarkable results have been achieved in water cut control and decline control. With the continuous deepening of development degree, new characteristics have emerged in water drive development, and the change law of indicators has also changed greatly. Therefore, it is necessary for us to find out the change law of fine potential tapping development indicators, so as to provide a theoretical basis for the next sustainable and high-efficiency development. Through theoretical calculation and numerical simulation, this study gives the law of production decline and water cut change in different stages before, during and after fine potential tapping; The laws of water cut and production decline in different stages are evaluated and predicted. It provides a theoretical basis for the fine potential tapping, popularization and application of other blocks.

Key words: Fine tapping potential; Development law; Water content change; Production decline.

1. Introduction

Block A has carried out fine potential tapping based on comprehensive utilization of well pattern and subdivision water injection, and achieved remarkable results in controlling water cut and decline. However, as the distribution of remaining oil is becoming more and more scattered, the adjustment potential is becoming smaller and smaller, and the adjustment is becoming more and more difficult. Therefore, the field test of water control and efficiency improvement focusing on increasing the production thickness of thin and poor layers and low efficiency circulation treatment has been carried out in block A, and the methods of further enhancing oil recovery and improving development benefits have been explored, and phased progress has been made. After more than two years of water control, efficiency improvement and potential tapping, block A has achieved a good potential tapping effect. However, compared with the fine potential tapping period, the development law will change. At present, there is a problem of unclear understanding of the development law of each stage. Although the fine potential tapping has achieved a good potential tapping effect, whether the change of water cut and decline indicators continues to be better or better, and whether there is any law to follow, we still need to improve our understanding. With the deepening of development, the adjustment difficulty of water drive development is increasing, the adjustment potential is decreasing, and the pertinence of tapping the potential is not strong. Therefore, it is urgent to find out the development law of fine tapping potential to provide technical support for the next development, popularization and application.

2. Study on the law of yield change

The output of oil and gas fields and oil and gas wells is not constant, but constantly changes with the process of oil and gas exploitation or the implementation of development measures. Oil and gas production may show an upward trend in a certain period, but may tend to be stable in other periods, but it develops and changes in a decreasing trend in most of the time of oil and gas field development. The practice of a large number of oil and gas field development shows that about half of the recoverable geological reserves are recovered in the process of production decline. During the decline period, not only the amount of oil and gas recovered is large, but also the duration is long, and the decline law is very complex. Therefore, Studying the law of production decline is of great significance to do a good job in oil and gas field performance prediction and oil and gas production planning. More importantly, only after understanding the law of production decline of oil and gas fields, can we take effective measures to prevent production decline in order to improve the final recovery of oil and gas fields.

2.1 Optimization of yield variation model

In order to verify the accuracy of the model, the primary well pattern with relatively stable decline without major

 https://doi.org/10.1051/e3sconf/202447801009

measures is selected for calculation. By comparing the predicted production value with the actual production value, the optimal model describing the production change of block A is finally determined. According to the fitted decline model formulas, the error between the predicted production and the actual production is calculated. The error of APS formula and logistics model is relatively small, but the above error is similar. Through the calculation of the difference between the predicted cumulative oil and the actual cumulative oil, it can be found that the error of Arps decline model is the smallest. To sum up, APS decline formula has small prediction error, simple operation and wide application. It is the optimal model for studying production decline in the experimental area (Table 1).

Table 1. Error calculation of prediction results

Decline model	Instantaneous production error	Cumulative oil production error
Arps	6.19%	-0.58%
Logistics	5.93%	-0.92%
Weibull	7.11%	-1.26%

2.2 Yield change law before fine tapping potential

Before fine potential tapping, the production will mainly rely on well pattern infilling to achieve production replacement. The basic well pattern corresponds to the development of main oil layers. After the initial production rise period, the production enters the decline stage. After the secondary infilling, the production decline slows down and fluctuates with the production of new wells. The output of block A has fluctuated for four times. First, due to the transfer of pumping in the whole region, the decreasing trend of production has been curbed, and the production has rebounded by a small margin; The reason is that the production of the first well increases and decreases again after the first well is put into production; The third fluctuation occurred during the secondary infill period. The production of secondary infill wells increased for a short time and reached the peak in 2000. However, since the production of secondary wells increased less than that of the other two sets of well patterns in 1998 and 1999, the whole region ushered in the fourth decline in 1998:; In 2007, there were three infills. The production of the three wells led to another increase in the production of the whole region. Due to the short production time, there was no significant decline.

Determine each decline time period of the whole area before fine potential tapping according to the production change curve of well pattern. According to the analysis process of Arps decline model above, analyze and calculate the phased decline of each set of well pattern before fine potential tapping through trial and error method for the whole area, basic well pattern, secondary well pattern, secondary and tertiary well pattern of block A (after three infilling, secondary and injection, and form a new well pattern with three times), Judge the production decline type of four sets of well patterns in stages. Before fine potential tapping, the overall test area follows hyperbolic decline. With the increase of grid connection density, the decline rate gradually decreases: the classified well pattern follows hyperbolic decline before fine potential tapping, and the decline rate is based on the well pattern < primary well < secondary and tertiary combination. The fitting results are shown in figures 1-4.

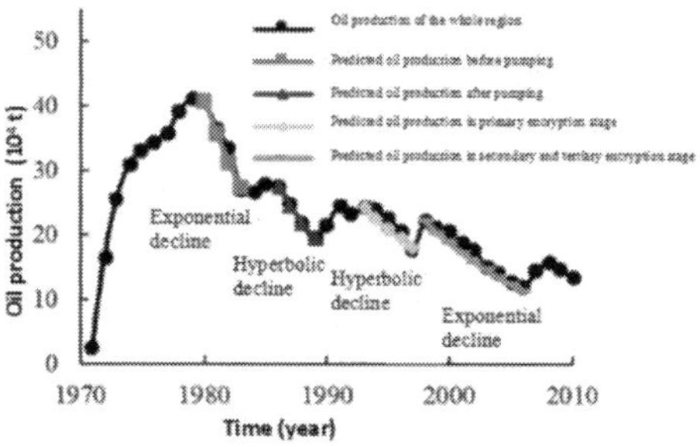

Figure 1 Oil production forecast of the whole region

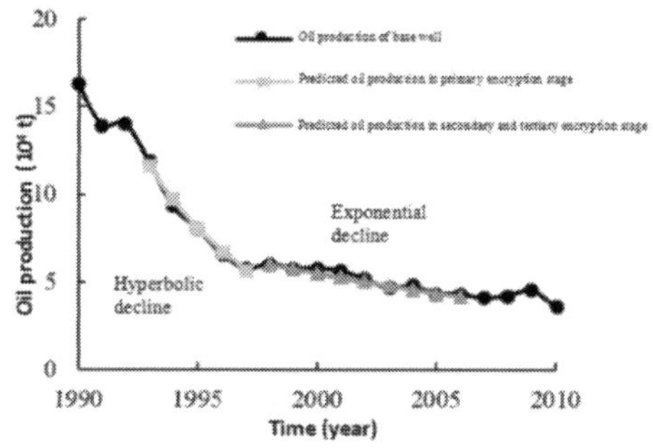

Figure 2 Oil production prediction of basic wells

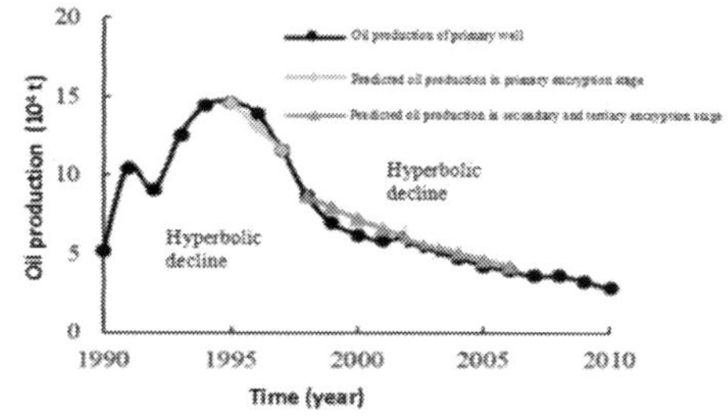

Figure 3 Oil production prediction of primarywells

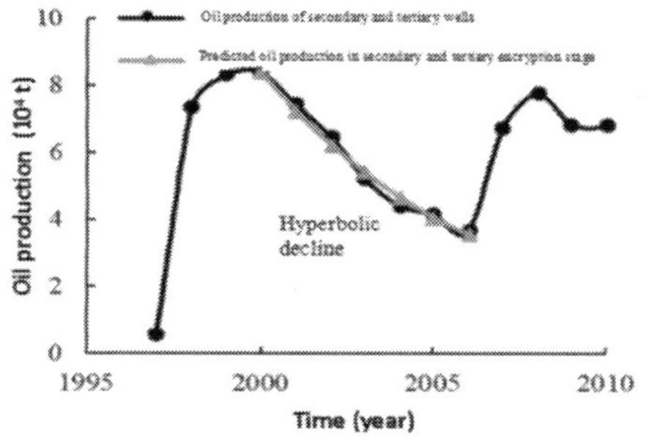

Figure 4 Oil production prediction of secondary and tertiary wells

2.3 Yield change law in fine tapping potential

The decline trend of output in the fine tapping stage is slower than that in the three infilling stage, and the output basically does not decline, with an average annual oil output of 133800 tons. In the early stage of the fine tapping stage, the oil production in the test area was affected by the oil increase measures, and the output maintained growth. When the effect of the measures gradually weakened, the output fell slightly in 2013. In 2014, the basic well pattern was greatly displaced and put into operation, and the output in the test area increased slightly compared with the previous year. The basic well pattern is in the decline stage at the initial stage, and the production increases after the production of extended reach wells; The primary well is less affected by the measures, and the production is in a steady decline state; In the initial stage of the second and third wells, the scale of measures is large, the oil increase is large, and the production increases. In the later stage, the effect of measures weakens and the production begins to decline

According to the change of grid connected production, determine the decline time periods of the whole region during and after fine potential tapping. According to the analysis process of Arps decline model above, calculate the phased decline of each set of well pattern in fine potential tapping stage by trial and error method. The fitting results are shown in figures 5-8.

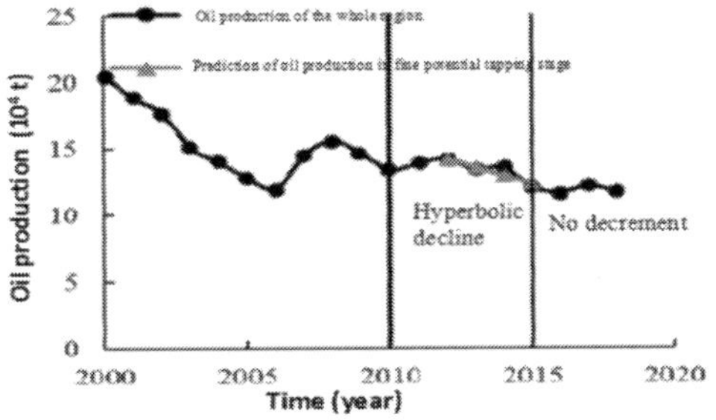

Figure 5 Oil production forecast of the whole region

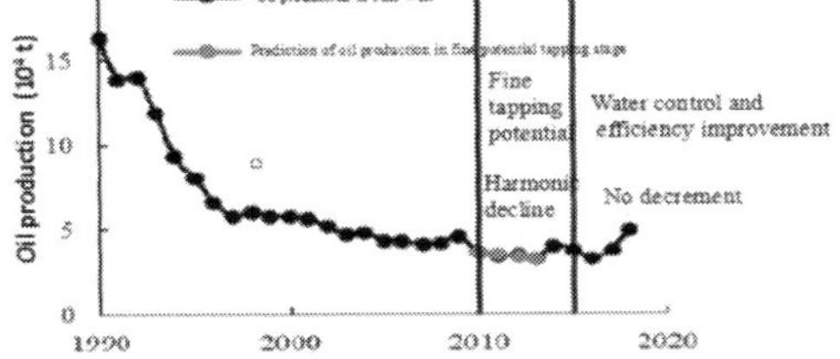

Figure 6 Oil production prediction of basic wells

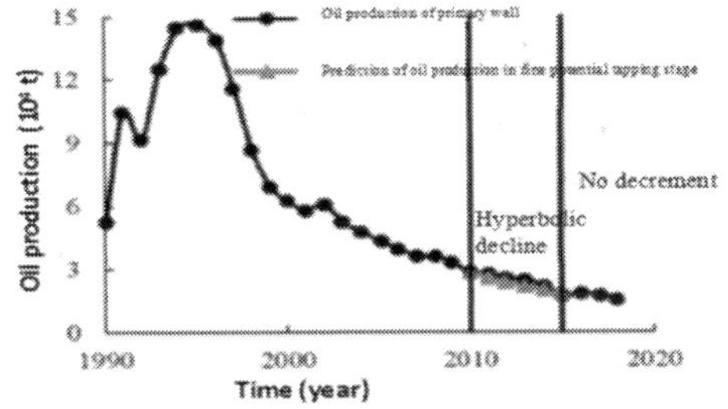

Figure 7 Oil production prediction of primary wells

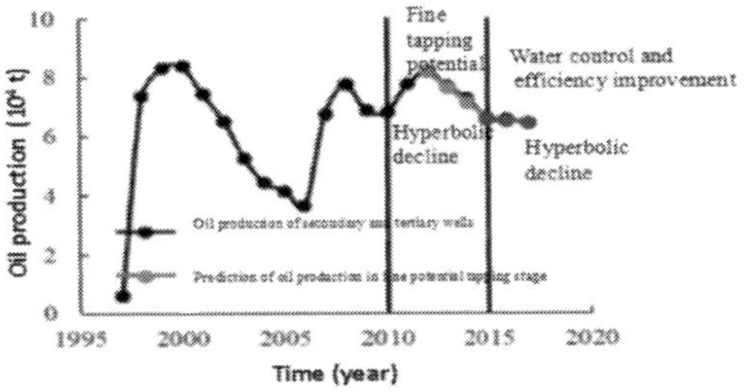

Figure 8 Oil production prediction of secondary and tertiary wells

In the stage of fine potential tapping, the decline type of well pattern changes, and the decline of the other two sets of well patterns slows down except the primary well pattern; In the stage of water control and efficiency improvement, there are only two or three times and obey the exponential decline.

Calculate the comprehensive decline rate and natural decline rate of each well pattern in the fine potential tapping stage respectively, and compare the difference of natural decline rate and comprehensive decline rate of each well pattern in the fine potential tapping stage.

There are obvious differences in natural decline and comprehensive decline among grid connected wells. The natural decline rate of basic wells < secondary and tertiary wells < primary wells, and the natural decline rate of block A in 2015 is 12.86%, while the comprehensive decline rate of secondary and tertiary wells < basic wells < primary wells, and the comprehensive decline rate of block A in 2015 is 4.55%

2.4 Yield change law after fine tapping potential

The natural decline and comprehensive decline of well pattern are calculated respectively, and compared with the average value in the last three years of fine potential tapping stage. The results are shown in the table below. It can be seen from the natural decline situation that the natural decline rate of block A continues to decrease in the stage of water control and efficiency improvement. At present, the decline rate is 6.54%, a decrease of 5.18% compared with the end of the previous stage. Among them, the decline rate of secondary and tertiary wells is large, and the natural decline rate in 2017 is only 2.86%; The natural decline rate of basic wells is higher than that of the previous stage. The main reason is that the extended reach wells enter the decline period and the decline rate is high, which makes the decline rate of basic well pattern higher than that of the previous stage; The natural production of primary wells increased in 2016 with a negative decline rate, but it resumed decline in 2017. The decline rate is equivalent to that of the previous stage, reaching 15.11%. The comprehensive decline of well pattern is similar to the natural situation. The comprehensive decline of the whole region continues to decrease, the decline of secondary and tertiary wells remains at a low level, the comprehensive decline of basic wells is slightly higher than that of the previous stage, but the decline rate is decreasing, and the comprehensive decline of primary wells has little change compared with that of the previous stage.

Comparing the difference between natural decline rate and comprehensive decline rate of well pattern in 2017, compared with the previous stage, the effect of measures is weakened, and the difference between natural decline and comprehensive decline is only 1.31%; No measures have been taken for the basic wells, and the natural decline rate is consistent with the comprehensive decline rate, both of which are 9.18%. The production decline rate of primary wells is the highest among the three sets of well patterns. The decline rate of secondary and tertiary wells is the smallest of the three well patterns. Compared with the previous stage, the scale of measures is reduced, and the difference between natural decline and comprehensive decline is reduced. However, thanks to the effect of measures, the decline rate continues to decrease.

2.5 Results of production decline law of well pattern in block A

According to the above Arps decline model analysis process, through the trial and error method for the basic well pattern, primary well pattern, secondary well pattern, secondary well pattern and tertiary well pattern in the whole area of block A, judge the production decline types of four sets of well patterns in stages, and calculate the variation law of the production of each set of well pattern with time and the variation law of the reduction rate with time. The calculation results show that: the whole obeys hyperbolic decline; With the increase of well pattern density, the decline rate decreases gradually; The classified well pattern follows hyperbolic decline before fine potential tapping, and the decline rate is based on the well pattern < primary well pattern < secondary and tertiary well pattern; In the stage of fine potential tapping, the decline type of well pattern changes. Except for the primary well pattern, the decline of the other two sets of well patterns slows down: in the stage of water control and efficiency improvement, only the secondary and tertiary well patterns obey exponential decline, but the decline rate is only 0.99%, and the production remains basically stable.

3. Study on variation law of water conten

The common characteristic of waterflooding oilfields is that the anhydrous oil production period is short, and there is a large amount of accompanying water production in the whole oil production process. However, the law of water cut rise is very different in different oilfields. Therefore, it is one of the important problems in the process of oilfield development to recognize the law of water cut change and take effective measures to reduce the accompanying water production.

3.1 Optimization of water cut fitting prediction model

Five types of displacement curves are used to fit the stage by stage water cut of the whole area and well pattern. Taking block A as an example, the fitting coefficients are compared first. Among them, the fitting correlation coefficients of convex and convex-s are higher, both reaching more than 0.95. On this basis, the error between the calculated predicted water cut value and the actual water cut is 2.83% and 3.72% respectively, Therefore, block A conforms to the convex water cut change. The water cut rises rapidly in the initial stage and slows down in the later stage. According to the fitted water cut model formulas, the error between the predicted water cut and the actual water cut is calculated. The prediction error of displacement series formulas is relatively small, simple and widely used. It is the best method to study the water cut prediction of block A (Figures 9-12).

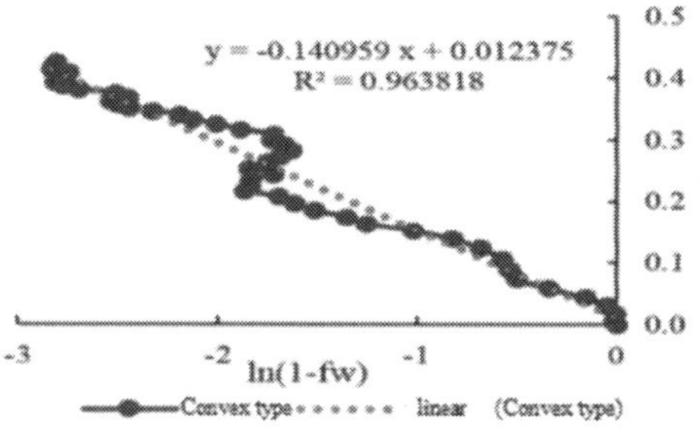

Figure 9 Convex water content change fitting curve

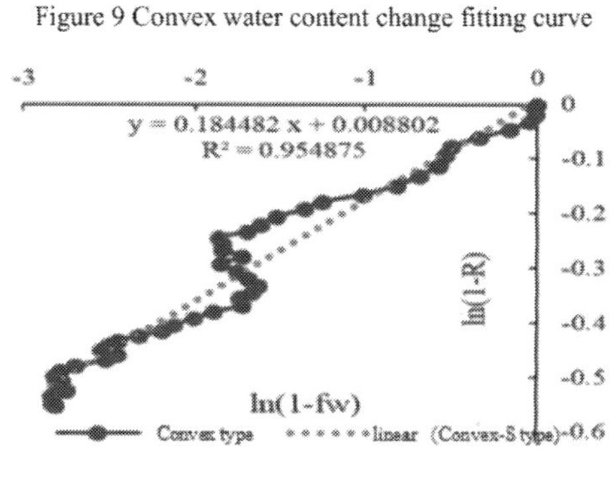

Figure 10. Fitting curve of convex-s type water cut change

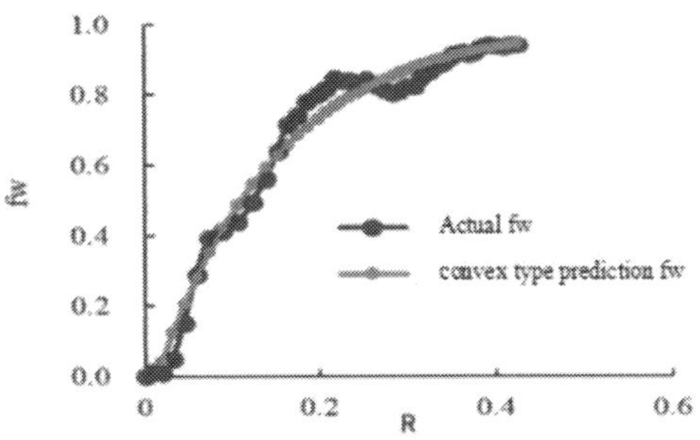

Figure 11. Convex predicted water cut curve

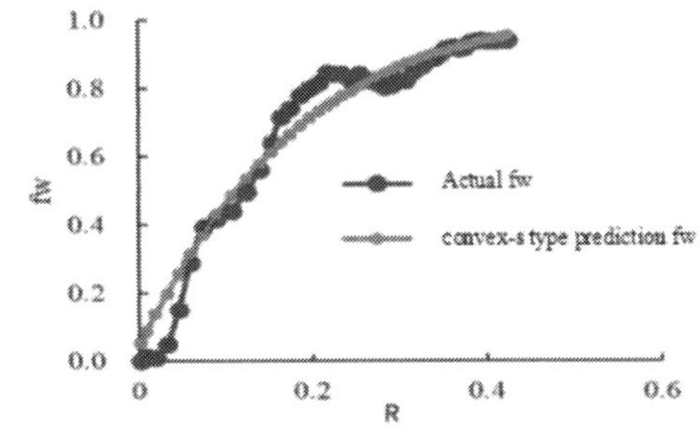

Figure 12. Convex-s type predicted water cut curve

3.2 Variation law of water content before fine potential tapping

The water free oil production period of the basic well before fine potential tapping is relatively short. After the initial production rise period, the production enters the decline stage, and the water cut continues to rise. In 1986, the water cut reached more than 80%. The water cut decline is mainly realized by well pattern infilling. In 1989, the primary well was put into operation and entered the primary infilling stage, and the water cut decreased to a great extent. The primary well experienced the water free period, and the water cut also continues to rise, In 1995, the water content of the primary well reached 77%, which was only 4% lower than that of the whole area. The secondary well was put into operation in 1996 and the tertiary well was put into operation in 2007. They did not experience the anhydrous oil production period, and the initial water content increased rapidly. In 2010, the water content of the whole area reached 93.8%, and the basic well, primary well and secondary and tertiary well were 95.5%, 94.4% and 91.7% respectively

The decline stage is determined by the previous oil production, and the decline of oil production will lead to the rise of water cut. Before fine potential tapping, the water cut is fitted by using the displacement series curve, and the water cut is predicted by the recovery degree. The water cut pattern of area a and sub well pattern is obtained. Before fine potential tapping, the whole area and sub well pattern obey convex and convex S-type water cut patterns as a whole, and the initial water cut rises rapidly.

3.3 Variation law of water content in fine potential tapping

In the stage of fine potential tapping, affected by the large-scale hole patching wells from 2010 to 2013, there were 236 wells in the region, 91 holes were patched, and the construction proportion reached 38.6%. The oil and liquid were increased in the test area, which greatly reduced the water content, and the change range of water content was very small. The water content in the region increased only 0.5% in five years: in order to exploit the non displaced crude oil at the edge of the fault, four large displacement wells were drilled at the end of 2013 and the beginning of 2014 to exploit the main oil layers, It is the same as the basic well, so it is classified as the basic well. It can be seen from the figure that the water cut of the foundation is greatly reduced due to the operation of the extended reach well.

During and after fine potential tapping (water control and efficiency improvement stage), except for the basic wells, the water content of block A and other well patterns has changed from the original convex and convex S-type to the current concave and concave S-type. The rising speed of water content has slowed down, and the measures have controlled the water content to a great extent.

3.4 Variation law of water content after fine potential tapping

After the fine potential tapping, in order to develop the crude oil of other faults, another 8 large displacement wells were drilled in 17 years, which greatly increased the oil production of the basic wells and the whole region, resulting in the continuous decline of water cut of the basic wells in 17 and 18 years; 16. The water cut did not rise in 17 years and decreased slightly. In 17 and 18 years, the water cut increased due to the impact of drilling shut in: the liquid production of secondary and tertiary wells accounted for 40% of the liquid production of the whole region. After the impact of drilling shut in, the water cut continued to rise. Generally speaking, the water cut was well controlled in the stage of water control and efficiency improvement

3.5 A block water cut mode determination results

According to the water content change curve of well pattern, determine the time period of each water content pattern in the whole area. According to the above water content pattern analysis process, predict the water content of the whole area, basic well pattern, primary well pattern, secondary and tertiary well pattern (after three times of infilling, the secondary well is transferred to injection, and a new well pattern is formed with the tertiary well), and judge the water content pattern of three sets of well patterns in stages, The variation law of the water content of each set of well pattern with time is calculated. The calculation results show that the overall water content of area a and sub well pattern follows the convex and convex S-type water content patterns. After fine potential tapping, except for the basic wells, the water content patterns of area a and sub well pattern change into concave water content curves, and the rising speed of water content slows down, indicating that the water content is controlled in a great situation after fine potential tapping, which has played a significant effect.

3.6 Variation law of water cut rise

The rising rate of water cut reflects the change of water cut in the oilfield. The smaller the rising rate of water cut, the better the development effect of the oilfield. In the practical application of the oilfield, the ratio of the difference between the water cut of the current year and the previous year and the difference between the recovery degree of the current year and the previous year is often used as the water cut rise rate of the oilfield in the current year. Draw the change curve of water cut rising rate with recovery degree and the change curve of water cut rising rate with water cut in block A (Figures 13-14).

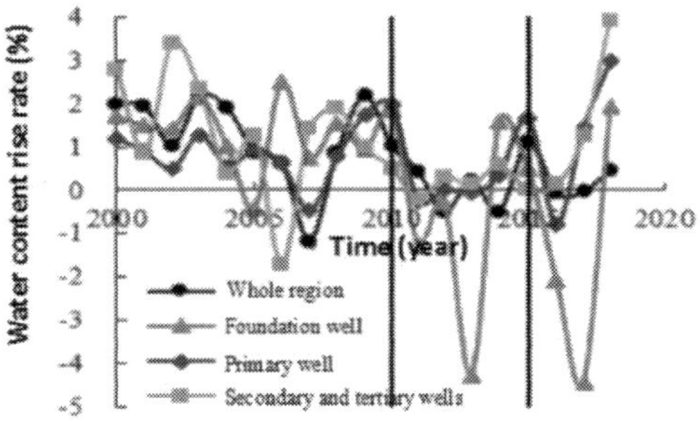

Figure 13. Curve of water content rise with time

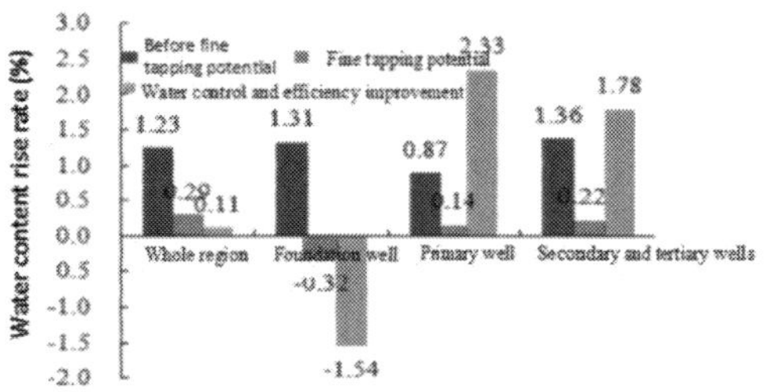

Figure 14. Phased statistical histogram of water content rise

It can be seen from the change curve of water content rise rate with time that after fine potential tapping, the trend of rapid water content rise has been effectively controlled. The phenomenon of water content rise rate less than 0 and water content decline has occurred in each year. The water content rise rate before fine potential tapping in the whole region is 1.23, the average water content rise rate in the stage of fine potential tapping is 0.29, the water content rise rate at the end of the stage is 0.11, and the average water content rise rate in the stage of water control and efficiency improvement is -0.04. In the stage of fine potential tapping, the rapid rise trend of water cut in the whole area and sub well pattern has been effectively controlled, and the maximum basic well pattern has been reduced by 1.63 percentage points. In the stage of water control and efficiency improvement, although the water cut rise rate of primary wells and secondary and tertiary wells has increased, the water cut rise rate of the whole region and basic wells has been controlled, and the maximum basic well pattern has been reduced by 1.22 percentage points.

3.7 Prediction results of water content without measures

In the third encryption stage, due to the rapid rise of water content, fine potential tapping measures were taken in area a, and hole patching was carried out from 2010 to 2013; After the completion of hole patching in 13 years, the water content of the whole area increased slightly after 14 years, and water control and efficiency improvement measures were taken in area a to stabilize the water content. If fine potential tapping is carried out at the end of 10 years and water control and efficiency improvement are not carried out in 16 years, the decline change of the whole region can be obtained, as shown in the figure below:

The water cut is predicted according to the previous decreasing oil production. The decreasing oil production is used to calculate the cumulative oil production NP, and the recovery degree R is calculated through NP to fit the displacement series curve. The predicted water cut without fine potential tapping conforms to the convex curve. The fitting parameters are determined to predict the future water cut. Assuming no fine potential tapping, the water cut will reach 97.1% in 2025, In the stage of fine potential tapping and water control and efficiency improvement, the rise of water content is controlled to a great extent; Because there are few water cut rising points without water control and efficiency improvement, the displacement series formula is no longer applicable. At this time, the future water cut is predicted through the C-type water drive law curve. The water cut without water control and efficiency improvement will reach 96.1% in 2025. The water cut is controlled to a certain extent in the stage of water control and efficiency improvement. These two prediction curves further illustrate that fine tapping potential has greatly reduced the rising trend of water cut, It can play the role of water control (Figure 15).

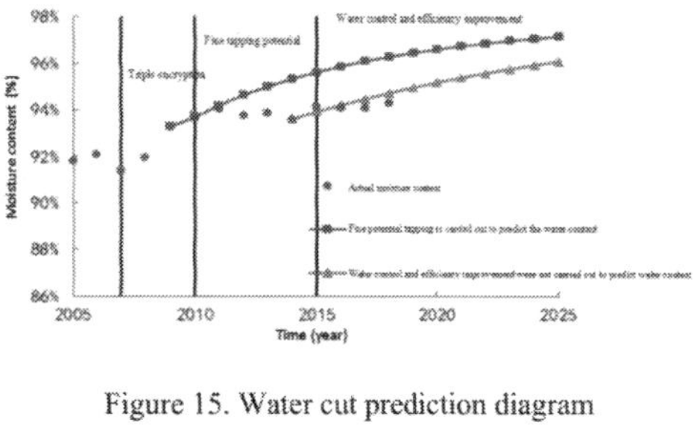

Figure 15. Water cut prediction diagram

4. Conclusion

1 Before fine potential tapping, the initial decline basic well pattern < primary well pattern < secondary and tertiary well pattern, the decline pattern in fine potential tapping changes from exponential decline to hyperbolic

decline, and the whole area conforms to exponential decline after fine potential tapping;
2 Before fine potential tapping, the whole area and sub well pattern obey convex and convex S-type water cut patterns as a whole. After fine potential tapping, except for the basic well pattern, the water cut patterns become concave water cut curves
3 After the fine potential tapping, the fast rising trend of water cut has been effectively controlled. In each year, the water cut rising rate is less than 0 and the water cut has decreased. In the fine potential tapping stage, the fast rising trend of water cut in the whole region and well pattern has been effectively controlled. In the stage after fine potential tapping, although the water cut rise rate of primary wells and secondary and tertiary wells has increased, the water cut rise rate of the whole region and basic wells has been controlled. Fine tapping the potential greatly reduces the rising trend of water content and can play a role in water control.

References

1. Yuan Qingfeng, understanding the law of oilfield development and developing oilfield scientifically and reasonably [J]. Daqing Petroleum Geology and development. 2004 (05)
2. Xu Jianjun, Huang Lida, Yan Limei, Yi Na. Insulator Self-Explosion Defect Detection Based on Hierarchical Multi-Task Deep Learning[J]. Transactions of China Electrotechnical Society, 2021, 36(07): 1407-1415.
3. Yan Limei, LIU Yongqiang, XU Jianjun,et al. Broken string diagnosis of composite insulator based on Grabcut segmentation and filler area discrimination[J].Power System Protection and Control, 2021, 49(22): 114-119
4. Yi, Q. Wang, L. Yan, et al., A multi-stage game model for the false data injection attack from attacker's perspective. Sustainable Energy Grids & Networks 28 (2021).
5. Na Yi,Jianjun Xu,Limei Yan,Lin Huang. Task Optimization and Scheduling of Distributed Cyber-physical System Based on Improved Ant Colony Algorithm. Future Generation Computer Systems, 109 (Aug. 2020),134-148.
6. Yang Zhao, Jianjun Xu, Jingchun Wu. A New Method for Bad Data Identification of Oilfield Power System Based on Enhanced Gravitational Search-Fuzzy C-Means Algorithm. IEEE Transactions on Industrial Informatics. VOL. 15, NO. 11, NOVEMBER 2019 5963-5970
7. Jing Han,Xi Wang,LiMei Yan, Aida Dahlak, et al. Modelling the performance of an SOEC by optimization of neural network with MPSO algorithm. International Journal of Hydrogen Energy, Volume 44, Issue 51, 22 October 2019, Pages 27947-27957.https://doi.org/10.1016/j.ijhydene.2019.09.055

Application of Seismic Attribute Analysis Technology Guided by Model Forward Modeling in L29 Area

Xiaodong Shi, Wei Zhang, Liquan Mao, Shougang Liu

Daqing Oilfield Co., Ltd. No. 9 Oil Production Plant, Daqing City, Heilongjiang Province, China

Abstract: Taking the SaErTU and Putaohua oil layers in the L29 well area of LHP oilfield in the northern Songliao Basin as the research object, In response to the difficulty of interference between thin interbedded sand and mudstone and strong reflection between strata, which have a significant impact on fine prediction of sand bodies, a stratigraphic model is established based on the geological characteristics of the target layer to eliminate the impact of stratigraphic reflection; Then add a thin layer of sand body to its interior, establish a thin interlayer model, and obtain a geological model that is more in line with the actual situation of the target layer. Using forward simulation, analyze the seismic response characteristics of sand bodies, extract 30 seismic attributes from 4 categories: amplitude statistics, composite seismic trace statistics, sequence and frequency spectrum statistics, and calculate the correlation between cumulative sandstone thickness and seismic attributes, and select a sensitive seismic attribute set;By combining sedimentary and drilling data, the seismic attribute with the highest sensitivity to the target layer is selected. The multi-level dimensionality reduction and gradual improvement of seismic attribute selection methods using "model forward modeling, attribute analysis, relevant optimization, and drilling implementation" can effectively improve the prediction accuracy of thin interbedded sand bodies and greatly reduce the risk of oilfield exploration and development.

Keywords: thin interbedded sand body; Seismic attributes; Forward simulation; Attribute optimization; Reservoir prediction.

1. Introduction

Seismic attributes are special measurements of geometric, kinematic, dynamic, or statistical characteristics derived from pre stack or post stack seismic data through mathematical transformations. The research and application of seismic attribute analysis technology play an important role in structural interpretation [1-2], reservoir prediction [3-5], and fluid detection [6-7]. But there are many types of seismic attributes, and different seismic attributes represent different meanings and play different roles in different fields. Some seismic attributes have clear physical meanings and are directly related to reservoir parameters or geological features, such as coherence, curvature, and ant body attributes, which mostly reflect fault characteristics [8]. However, most seismic attributes do not have a clear physical meaning, only mathematical operations on seismic amplitude, frequency, and phase. It is necessary to analyze the geological significance represented by seismic attributes based on actual geological conditions. Optimizing seismic attributes or combinations of seismic attributes is a key issue in achieving reservoir prediction [9]. Many scholars at home and abroad have conducted research on the prediction of thin interbedded sand bodies, and have achieved certain results. An Peng et al. [10] On the basis of analyzing the tuning effect, time-frequency analysis technology is used to optimize the sensitive frequency, and the data volume is reconstructed based on the sensitive frequency to eliminate similar single sand body thicknesses but significant differences in seismic waveform response characteristics. The sensitive seismic attributes are optimized to predict the distribution of sand bodies. Zhang Junhua et al. [11] used 90 ° phase shift technology to improve the well seismic relationship and effectively improve the accuracy of reservoir prediction for thin interbedded reservoirs. Peng Zuolei et al. [12] used principal component analysis to perform principal component analysis on various seismic attributes, establish the connection between multiple seismic attributes and thin reservoirs, reduce the multiplicity of single seismic attribute analysis, and use support vector machine method to predict thin reservoirs. Liu Wei et al. [13] established a typical thin interbedded sand and mudstone model, focusing on analyzing the variation characteristics of 8 seismic attributes, and summarized their variation patterns with the thickness of thin interbedded sandstone, providing a certain theoretical basis for optimizing the prediction attributes of thin interbedded sand and mudstone reservoirs.

From the literature, scholars in the past mainly focused on the study of the interference effect between thin interbedded sand bodies, and there was little research and practical analysis on the impact of strong reflection between formations on reservoir prediction. This article takes the SaErTU and Putaohua oil layers in the L29 well area of L oilfield as examples. Based on the geological characteristics of the target layer and the development of the sand body, the influence of stratigraphic reflection is first considered to establish a stratigraphic model, and then a thin layer of sand body is added to its interior to obtain a more practical geological model. Afterwards, seismic forward modeling was conducted to calculate the correlation between cumulative sandstone thickness and seismic attributes, and a sensitive seismic attribute set was selected. By combining sedimentary and drilling data, the seismic attributes with the highest sensitivity to the target layer are selected to improve the prediction accuracy of thin interbedded reservoirs.

2. Geological Overview of the Work Area

The L29 area is located on the eastern slope of the L anticline in the northern part of the Songliao Basin. The Sa and Pu oil layers in this area are controlled by the Qiqihar sedimentary system in the northwest and the Nehe Yi'an sedimentary system in the north. The delta front channel sand bodies and sheet sand bodies are vertically staggered and severely heterogeneous, making them typical thin interbedded reservoirs (Figure 1). The interference between sand bodies has a significant impact, making it difficult to predict advantageous sand bodies.

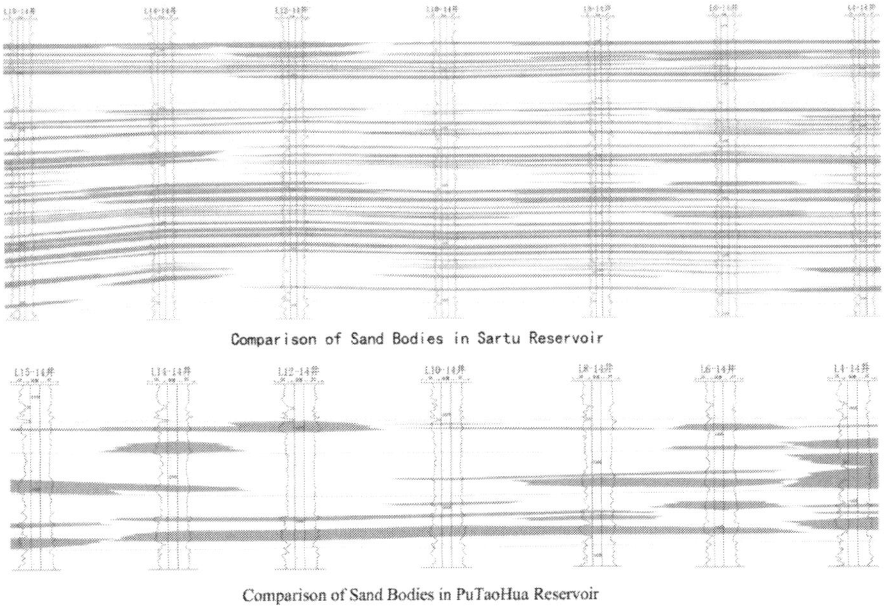

Figure 1. L29 area Comparison of Sand Bodies in the Development Zone of Sartu and Putaohua Oil Layers

3. Optimization of forward modeling attributes for earthquake models

3.1 Wedge model

As shown in Figure 2, a set of wedge-shaped body models is designed, with the thickness of the wedge-shaped body gradually increasing from 0m to 40m from left to right. The proportion of internal sandstone is 25%, 50%, 75%, and 100%, respectively. In the model, yellow represents sandstone, with a velocity of 4650m/s and a density of 2. 5g/cm3; White represents the surrounding rock, with a velocity of 3800m/s and a density of 2. 4g/cm3. The seismic reflection record is obtained by using a 45 Hz main frequency Reyker wavelet and a self excited and self received ray tracing method.

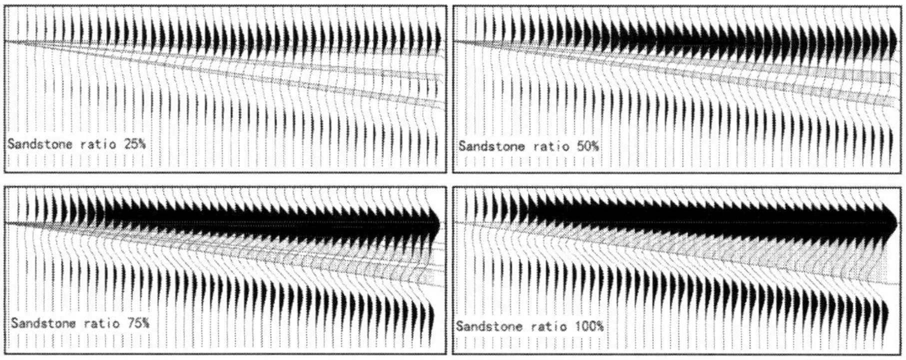

Figure 2. Wedge shaped thin interbedded seismic forward modeling

Extract the amplitude and frequency of each wave peak on the top surface of the wedge-shaped body model, and draw the amplitude and frequency tuning curve (Figure 3). The amplitude tuning curve indicates that regardless of how the sandstone ratio changes, the amplitude reaches its maximum near the tuning thickness. But as the sandstone content decreases, the tuning thickness decreases, the amplitude weakens, and the intensity changes become more complex. The frequency tuning curve shows that the thickness of the wedge-shaped body is below 27m, and the frequency decreases with the increase of thickness. The change in sandstone content has little effect on the frequency attribute; When the thickness is greater than 27m, the frequency begins to increase. The sandstone content has a significant impact on the frequency attribute. The smaller the sandstone content, the greater the rate of frequency increase.

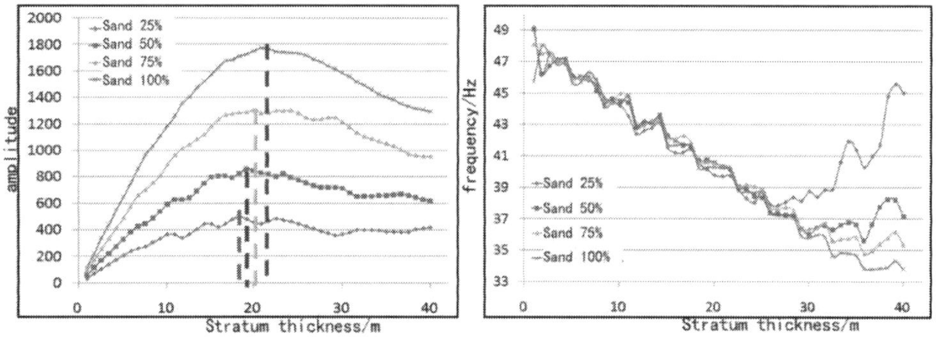

Figure 3. Wedge body model amplitude and frequency tuning curve diagram

The forward modeling of the thin interbedded sand and mudstone model (Figure 2 and Figure 3) shows that both changes in sandstone content and thickness can cause significant changes in seismic amplitude. The amplitude attribute can be used to predict the development of sandstone [14-16]. However, in practical applications, it has been found that the thin interbedded model did not take into account the impact of strong reflections between strata on the reflection shielding of sand bodies. As a result, starting from this forward model, it can be found that multiple seismic attributes have good sensitivity to sand bodies and high consistency. However, the predictive ability of the seismic attribute sand bodies extracted in practice varies greatly. It is difficult to optimize the seismic attributes that are most sensitive to sand bodies using this type of model.

3.2 Forward modeling of thin interbedded sand and mudstone based on stratigraphic models

Thin interbedding refers to a set of strata composed of multiple thin layers that cannot be distinguished from each single layer by seismic reflection. Thin interbedded reflection wave is a composite wave formed by the interference of multiple thin layer reflections, which cannot distinguish each single layer. Therefore, the thin interbedded reflection wave is the overall seismic response of the thin interbedded layer [17]. The impact of strong reflections between superimposed strata on the reflection shielding of thin interbedded sand bodies has brought significant difficulties to the prediction of thin interbedded sand bodies. The accuracy of seismic attribute prediction results is often low and the multiplicity of solutions is strong. Conventional thin interlayer models are difficult to characterize the seismic response characteristics of the target layer and cannot be used for seismic optimization. In order to improve prediction accuracy and reduce multiplicity, a geological model is first established to eliminate the strong reflection shielding effect between strata, and then sand bodies are added to it to establish a thin interbedded model of sand and mudstone. The results obtained using this geological model will be more targeted and applicable.

Analyze the drilling data in the area, determine the velocity and density parameters of surrounding rock, sand, and mudstone in each formation (Table 1), and design stratigraphic models for the Sartu and Putaohua oil layers based on geological characteristics. Obtain forward seismic reflection records using the self excited and self collected ray tracing method. As shown in Figure 4, the forward modeling results of the formation are as follows: strong amplitude peak reflection on the top surface of the S II oil layer group, medium strong amplitude peak reflection on the S I top, S III top, and P top surfaces, while medium strong amplitude valley reflection is formed at the bottom boundary of P due to contact with the underlying low-speed formation. The seismic forward model is consistent with the reflection characteristics of the actual seismic profile, indicating that the model parameters are relatively reasonable. Ultimately, the sand mudstone thin interbedded model can effectively eliminate the influence of interlayer reflection.

Table 1 Forward Modeling Parameters of Each Layer Group

Forward layer group		Interval transit time(us/m)	Velocity(m/s)	Density(g/cm^3)	Impedance (m/s*g/cm^3)
SITop roof		374	2673. 8	2. 42	6470. 6
SI	mudstone	352	2840. 9	2. 44	6931. 8
	sandstone	258	3876. 0	2. 58	10000. 0
SII	mudstone	298	3355. 7	2. 51	8422. 8
	sandstone	209	4784. 7	2. 5	11961. 7
SIII	mudstone	290	3448. 3	2. 54	8758. 6
	sandstone	209	4784. 7	2. 5	11961. 7
P	mudstone	280	3571. 4	2. 59	9250. 0
	sandstone	232	4310. 3	2. 51	10819. 0
P Lower roof		303	3300. 3	2. 6	8580. 9

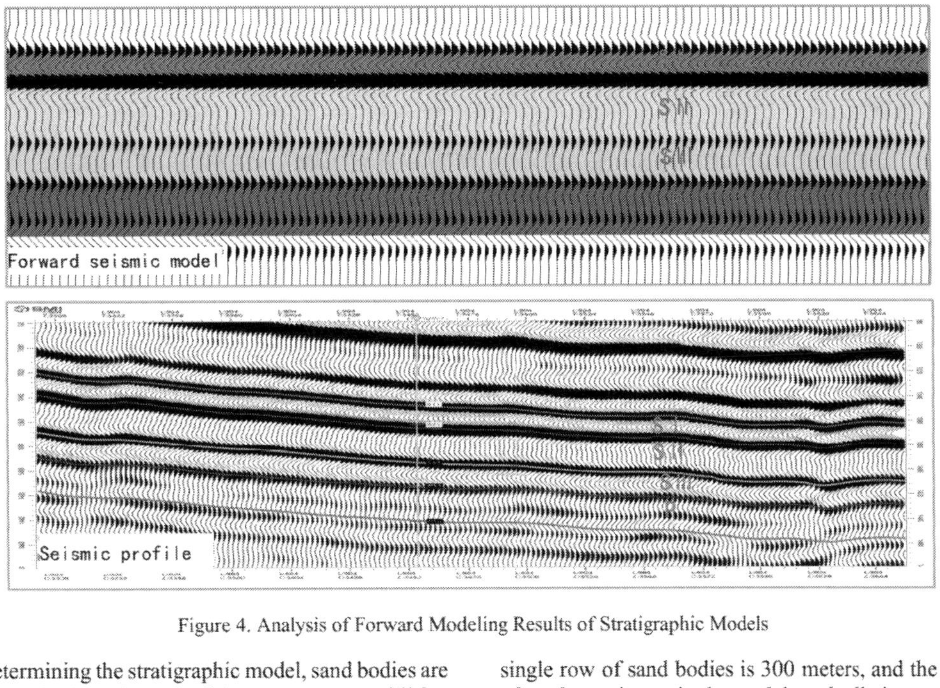

Figure 4. Analysis of Forward Modeling Results of Stratigraphic Models

After determining the stratigraphic model, sand bodies are added layer by layer in each oil layer group to establish a thin interbedded model of sand and mudstone. As shown in Figure 5, taking the S I oil reservoir group as an example, the thickness of a single layer of sandstone is 2 meters, with an interval of 1 meter. The width of each single row of sand bodies is 300 meters, and the number of sandstone layers in the model gradually increases from 1 to 7 from left to right. Forward seismic reflection records were obtained using the self excited and self received ray tracing method using 45Hz Reke wavelet.

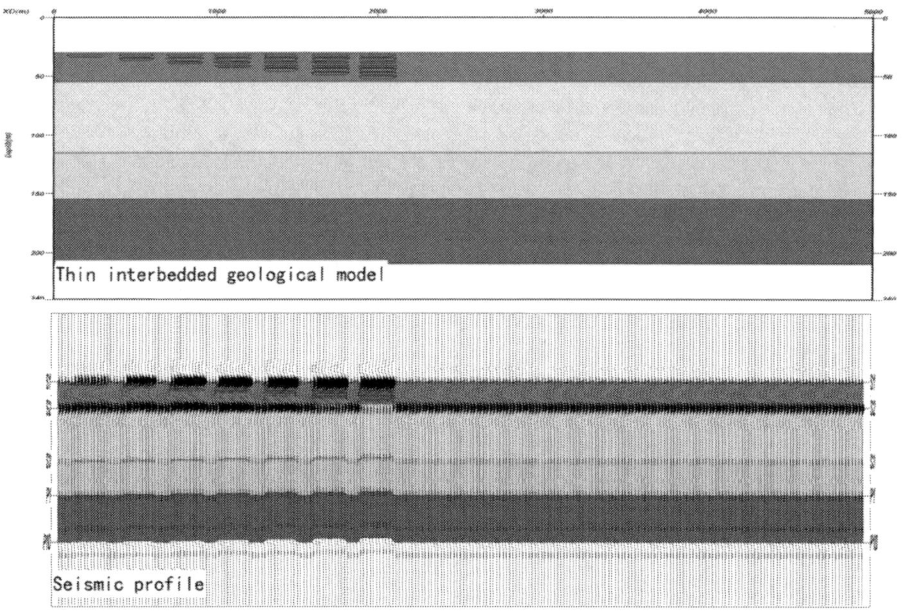

Figure 5 Geological model and seismic forward modeling of thin interbedded layers in S-I oil reservoir group

From the seismic forward modeling, it can be seen that as the thickness of sandstone increases, the amplitude shows a significant increasing trend. Although the number and total thickness of thin interbedded layers of sand and mudstone in these 7 columns are different, their seismic reflection characteristics are roughly the same, that is, the reflection waveforms are all composite waves with a single waveform, and there is no increase in the intermediate reflection interface compared to the stratigraphic reflection.

3.3 Seismic attribute optimization

For the seismic forward modeling of thin interbedded sand and mudstone in the S I oil layer group, a total of 30 seismic attributes were extracted from the top and bottom interfaces of the oil layer group, including amplitude statistics, composite seismic trace statistics, sequence and spectrum statistics, and a graph of the variation of seismic attributes with cumulative sandstone thickness was drawn. As shown in Figure 6, the horizontal axis in the figure represents the forward seismic channel number, the right vertical axis represents the cumulative thickness of sandstone, and the black line represents the thickness values of sandstone for different channels. It can be seen that as the channel number increases, the cumulative sandstone thickness increases, and the size of the channel number can represent the size of the cumulative sandstone thickness. The left vertical axis represents the normalized seismic attribute values, and the colored lines represent the variation curves of various seismic attributes with the cumulative thickness of sandstone. If a certain seismic attribute value monotonically increases or decreases as the channel number increases, it indicates a good correlation with the cumulative sandstone thickness and can be used to predict the development of sandstone.

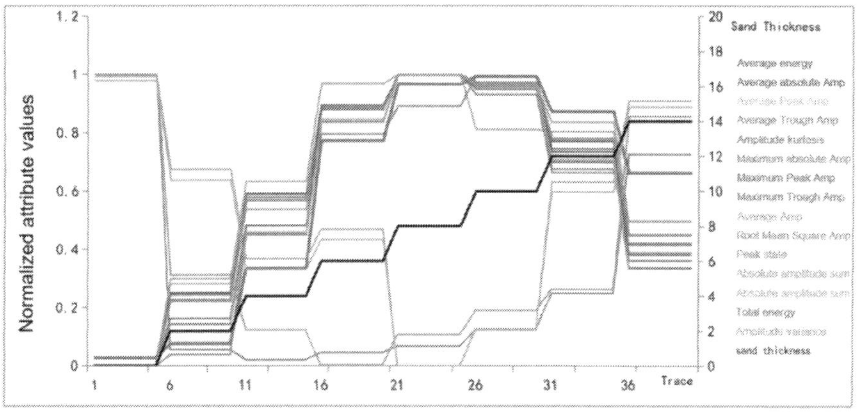

Figure 6 Amplitude statistical seismic attributes versus cumulative sandstone thickness variation

For example, the maximum peak amplitude value in Figure 6 increases with the increase of cumulative sandstone thickness, which has a good correlation with sandstone thickness. It reaches its maximum value at a thickness of 10m, and then the maximum amplitude value slightly decreases. Due to the average development of sandstone in the S I oil reservoir group, the accumulated maximum thickness is only about 10m after drilling, so the maximum peak amplitude can be used to predict the development of sandstone. Among the amplitude statistical attributes, the root mean square amplitude attribute has a good correlation with sandstone thickness and can also be used to predict the development of sandstone. As shown in Figure 7, in the statistical category of composite seismic traces, there is a negative correlation between the average instantaneous frequency attribute and the cumulative sandstone thickness. As the sandstone thickness increases, the attribute values monotonically decrease. The two have a good correlation and can be used to predict the development of sandstone. In the sequence and spectrum statistics category (Figure 8), the correlation between various seismic attributes and cumulative sandstone thickness is generally slightly worse than that of amplitude statistics and composite seismic trace statistics, and their seismic attributes are not suitable for predicting sandstone in the S I oil layer group. Therefore, based on the statistical analysis of the thin interbed forward model of the stratigraphic model, a seismic attribute set with maximum peak amplitude, root mean square amplitude, and average instantaneous frequency is preliminarily selected.

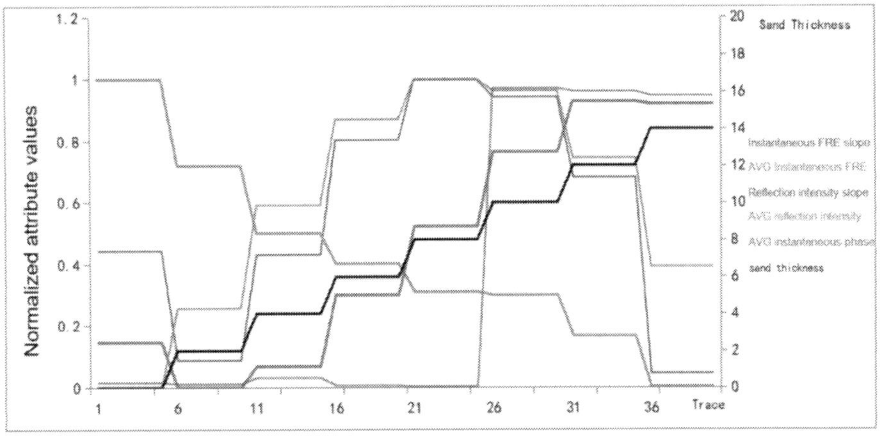

Figure 7. Changes in cumulative sandstone thickness due to statistical seismic attributes of composite seismic

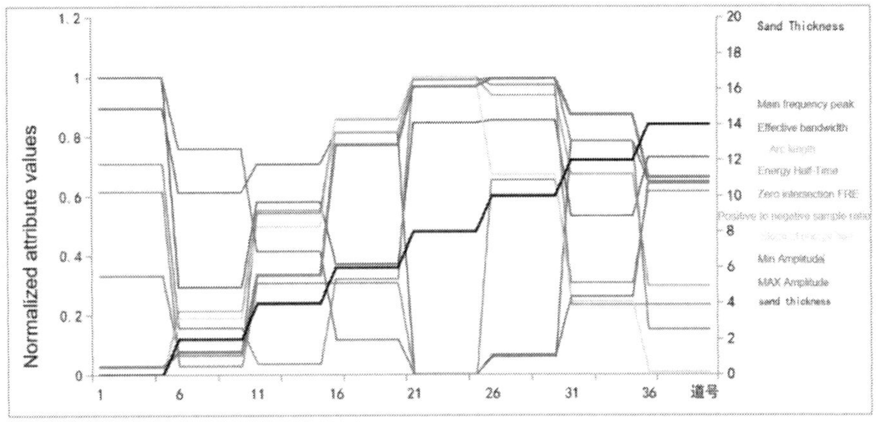

Figure 8. Changes in cumulative sandstone thickness due to seismic attributes of sequence and spectral statistics

Based on the preliminary determination of sensitive attributes, further attribute optimization is carried out based on regional sedimentary characteristics and drilling data. As shown in Figure 9, based on the development of sandstone in the S1 oil layer group, it is believed that a cumulative sandstone thickness of over 3m is the advantageous reservoir for this oil layer group. Therefore, the seismic attribute prediction of the oil layer group is adjusted to yellow and red for areas with a cumulative sandstone thickness of more than 3m, and to blue for areas with a cumulative sandstone thickness of less than 3m. Calculate the prediction accuracy of each sensitive seismic attribute, with a maximum peak amplitude prediction accuracy of 81. 3%, a root mean square amplitude prediction accuracy of 81. 3%, and the average instantaneous frequency prediction accuracy of 87. 5%. During the sedimentation period of the S I oil layer formation, as the lake level rises, the delta gradually retreats northward, with a large area of inter flow bay facies distributed in the southern part, and shore-shallow lake facies visible in the southeast. The average instantaneous frequency attribute can better characterize this sedimentary feature. Therefore, the average instantaneous frequency attribute is ultimately selected to characterize the development of sandstone in the S I oil reservoir group. The above methods can be used to optimize the seismic attributes that are most sensitive to sandstone and conform to sedimentary characteristics.

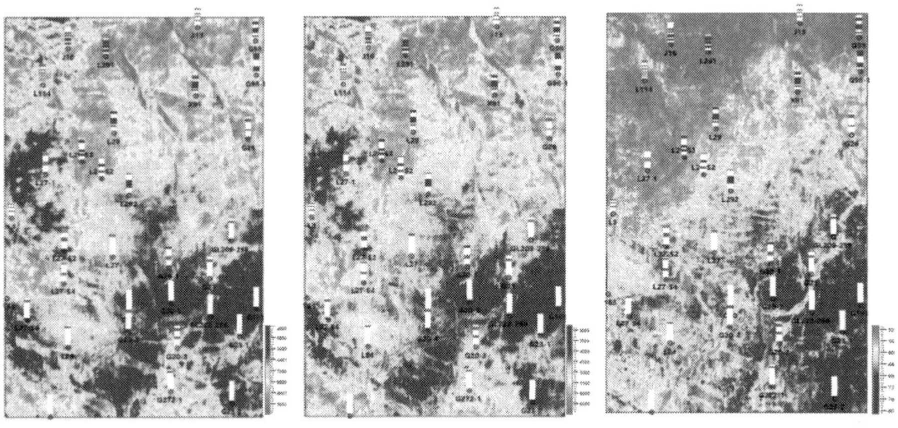

Figure 9. Preliminary Selection of Seismic Attribute Set for S-I

4. Application Effect Analysis

Adopting the forward seismic attribute optimization method based on the stratigraphic model for the thin interbed model, the seismic attribute optimization was carried out for the three oil layer groups of the Sartu oil layer and the three sandstone groups of the Putaohua oil layer. During the sedimentation period of the Sartu oil layer, it was mainly composed of sheet-like sand deposits, with sand bodies distributed in a sheet-like manner, good continuity, and relatively large scale. Figure 10 shows the seismic attribute maps selected for the three oil layer groups in Sartu, with an average instantaneous frequency prediction accuracy of 87. 5% for the S-I oil layer group, an average amplitude attribute prediction accuracy of 84. 4% for the S-II oil layer group, and an reflection intensity slope prediction accuracy of 81. 3% for the S-III oil layer group. The attribute prediction shows the development of sandstone in each oil layer group, with a flaky distribution and good continuity, which can better characterize the sedimentary characteristics of each oil layer group in the area and effectively predict the sandstone development area.

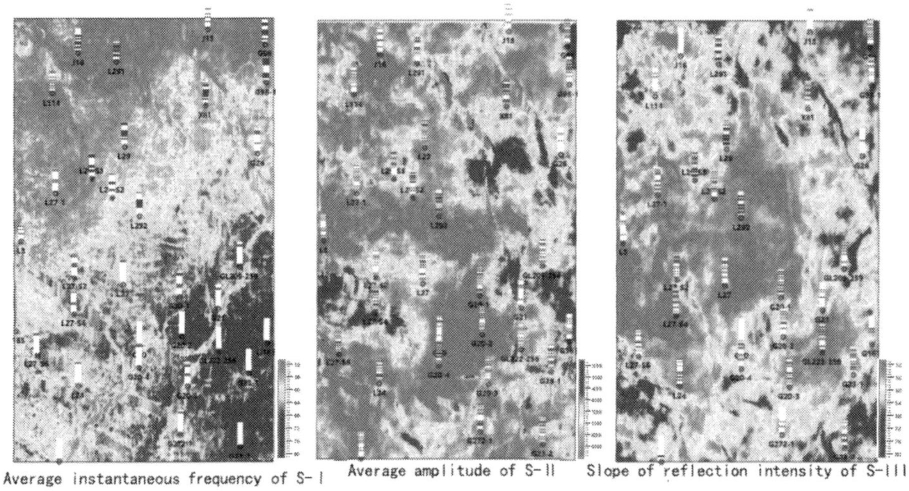

Figure 10. Optimization Results of Properties of Various Oil Layer Groups in Sartu Reservoir

During the sedimentation period of the Putaohua oil layer, there was a sedimentary feature of water intrusion from bottom to top. The underwater distributary channels in the delta front are frequently diverted and vertically distributed in staggered and continuous layers. On the plane, the lateral changes of the reservoir are rapid, and the sand bodies are mainly distributed in strip or lens shapes, with short extension distance and small development scale. Adopting the forward seismic attribute optimization method based on the stratigraphic model, the root mean square amplitude attribute of the upper sand formation of the Putaohua oil layer is optimized, with a prediction accuracy of 81. 3%. The minimum amplitude attribute is selected for the sand group in the grape blossom, with a prediction accuracy of 84. 4%. The optimal average absolute amplitude attribute for the lower sand formation of the Putaohua oil layer is selected, with a prediction accuracy of 84. 4%. The selected seismic attribute predictions have a coincidence rate of over 80% and can better characterize the sedimentary characteristics of various sandstone formations in the Putaohua oil layer.

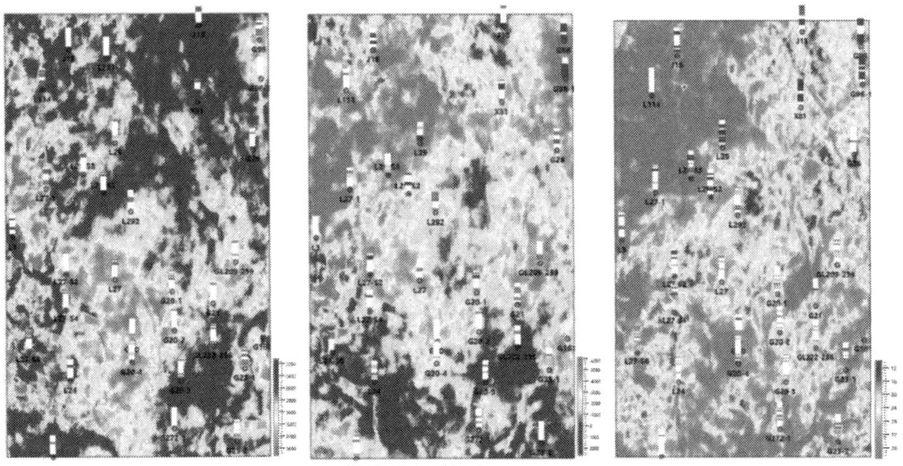

Figure 11 Optimization Results of Properties of Each Sandstone Formation in Putaohua Reservoir

5. Conclusion and Understanding

1) The forward modeling of the thin interbedded wedge-shaped body model shows that both changes in sandstone content and thickness can cause significant changes in seismic amplitude. The amplitude attribute can be used to predict the development of sandstone. However, in practical applications, it has been found that the thin interbedded model does not take into account the impact of strong reflections between strata on the reflection shielding of sand bodies, making it difficult to select the most sensitive seismic attributes for sand bodies.

The multi-level dimensionality reduction and gradual improvement of seismic attribute selection methods using "model forward modeling, attribute analysis, relevant optimization, and drilling implementation" can effectively improve the prediction accuracy of thin interbedded sand bodies. This greatly reduces the problem of multiple solutions in predicting seismic attributes of thin interbedded sand bodies, and provides an effective method for optimizing seismic attributes of thin interbedded reservoirs.

References

1. Lei Chen, Jianjun Xu, Shuang Chen, Hui Yang, Linhu Liu. Power Quality Disturbance Signal Denoising Based on Overcomplete Representation. IEEJ TRANSACTIONS ON ELECTRICAL AND ELECTRONIC ENGINEERING.2022; 17: 544–555
2. Xu Jianjun, Huang Lida, Yan Limei, Yi Na. Insulator Self-Explosion Defect Detection Based on Hierarchical Multi-Task Deep Learning. Transactions of China Electrotechnical Society, 2021, 36(07):1407-1415.
3. Yan Limei, LIU Yongqiang, XU Jianjun, et al. Broken string diagnosis of composite insulator based on Grabcut segmentation and filler area discrimination. Power System Protection and Control, 2021, 49(22): 114-119
4. Yi, Q. Wang, L. Yan, et al., A multi-stage game model for the false data injection attack from attacker's perspective. Sustainable Energy Grids & Networks 28 (2021).
5. Na Yi, Jianjun Xu, Limei Yan, Lin Huang. Task Optimization and Scheduling of Distributed Cyber-

physical System Based on Improved Ant Colony Algorithm. Future Generation Computer Systems, 109(Aug. 2020),134-148.

6. Yang Zhao, Jianjun Xu, Jingchun Wu. A New Method for Bad Data Identification of Oilfield Power System Based on Enhanced Gravitational Search-Fuzzy C-Means Algorithm. IEEE Transactions on Industrial Informatics. VOL. 15, NO. 11, NOVEMBER 2019 5963-5970
7. Xiong Xiaojun, Yin Cheng, Zhang Bailin et al. Method research of forecasting oil and gas using higher-order statistics. Chinese Journal of Geophysics, 2004, 47(5): 920-927.
8. Wang Bin, Zhu Shuairun, Wang Liang, et al. Prediction of channel sands based on seismic attribute analysis[J]. Petroleum Geophysics, 2019, 17(2): 45-49.
9. Jing Yongquan, Luan Dongxiao, Zhang Yuqing, et al. Prediction method for fluvial superimposed sandstone reservoirs based on seismic attribute characteristics[J]. Oil Geophysical Prospecting, 2018, 53(5): 1049-1058.
10. An Peng, Yu Zhilong, Liu Zhuan, et al. The application of sensitive frequency seismic attributes to thin sand body prediction: Exemplified by Zhaoyuan area in Songliao Basin[J]. Geophysical and Geochemical Exploration, 2020, 44(2): 321-328.
11. Zhang Junhua, Chang Jiangqiang, Wang Zuoqian, et al. Theoretical annotation and application of 90° phase-shifting thin interbed interpretation technique[J]. Journal of Jilin University (Earth Science Edition), 2022, 52(4): 1348-1359.
12. Peng Zuolei, Cheng Ming. Multivariate seismic attribute analysis improve prediction accuracy of thin reservoirs[J]. Fault-Block Oil & Gas Field, 2020, 27(3): 318-322.
13. Liu Wei, Yin Cheng, Wang Min, et al. Study on the characteristics of basic seismic attributes in fluvial sand-shale interbedded layers. Geophysical Prospecting for Petroleum, 2014, 53(4):468-476.
14. Zhang Mingxue, Cheng Xiaolei, Yu Tong. Seismic attribute optimization methods research based on forward simulation[J]. Journal of Heilongjiang University of Science & Technology, 2017, 27(6): 626-631.
15. Li Wei, Yue Dali, Hu Guangyi, et al. Optimization of seismic attributes in frequency bands and prediction methods for sand bodies - a case study of the northern area of Qinhuangdao 32-6 oilfield[J]. Oil Geophysical Prospecting, 2017, 52(1): 121-130.
16. Wang Yanguang, Li Hao, Li Guofa, et al. A seismic attribute compliant method for estimating the thickness of thin and thin interbedded sand bodies[J]. Oil Geophysical Prospecting, 2020, 55(1): 153-160.
17. Huang Wenfeng, Yao Fengchang, Li Hongbing. Regularities of tuning effects of thin interbedded layers and their net thickness determination[J]. Oil Geophysical Prospecting, 2012, 47(4):584-591.

The method of identification of abandoned channel by dense well pattern and its application in A Oilfield

Xiaolei Wang

Reservoir evaluation room of Geological Research Institute, the fifth oil production plant, Daqing Oilfield, China

Abstract: The existence of abandoned channel represents the end of the evolution history of single river channel. Taking the C layer in B area, A oilfield as an example, synthesizing the logging response characteristics, plane distribution pattern and the relative depth map method of sand top of abandoned channels, abandoned channels were identified under the guidance of modern deposition theory. In the three-dimensional window, a multi-angle observation and analysis of connected well profiles was carried out by using the underground dense well pattern data, and a total of 10 parts were identified in the research area, including s-type and pen-moon abandoned channel.

Key words: Meandering stream; abandoned channel; Point bar; Reservoir architecture.

1. Introduction

The C layer in B area belongs to distributary channel deposit in delta plain, where channel bodies of meandering stream are developed. The oil field has entered the high water cut stage, and the reservoir heterogeneity has a great influence on reservoir development. The remaining oil is highly dispersed. abandoned channel is one of the important factors affecting the connectivity of sand body. Due to the difference of lithology and physical properties, the upside of abandoned channel with silt and mud forms an effective occlusion among its surrounding point bars. Therefore, the identification of abandoned channels is of great significance to the study of reservoir heterogeneity and residual oil distribution.

2. The method of identifying abandoned channel in dense well pattern

There are three types of occlusion of abandoned channels, namely: abrupt abandonment, gradual abandonment, transition.

The roof of the dam separate abrupt abandonment channel from the new channel has been higher than the water level of the mainstream river, preventing the water from entering the abandoned channel and making the two completely isolated. Only during the flood season can the floodwater carry fine suspended matter into the abandoned channel. In the still water environment, the deposit is mainly mud-based and silty. finally, a circular enclosed lithologic blocking zone dominated by mud-texture strip is formed on the plane.

The roof of the dam separating the gradually abandoned channel is lower than the water level of the mainstream river. It has been at a low water level. Its internal water body always connects with upper water body of mainstream river channel. Most of the upper reaches flow downstream through new river channel (mainstream channel), and a small number of water will pass through the abandoned channel. The water in the abandoned channel is weak. Due to the semi-occlusion of the dam, middle coarse materials are difficult to move into the abandoned channel, which is dominated by rolling and jumping components at the bottom of the water body of the main river. Abandoned channel constantly receives fine particles deposited from the main river, which are mainly suspended and mixed with a small number of jumping components. Therefore, the sediment particle size of the abandoned channel is smaller than that of the surrounding point bar, and the shale content is higher.

Finally, The physical-lithology shielding strip is formed on the plane, which is dominated by physical property.

The transitional abandoned channel has both characteristics and forms the "walled type" occlusion of physical-lithology characteristics.

The above three different types of abandoned channels deposit, form a circular enclosed walled shielding strip on the plane, which separates contiguous meander sand bodies into combination of multiple point bars. The sediment of abandoned channel can play a certain protective effect on the fluid in the sand of the point bars on both sides. Its shielding intensity can be described by the ratio of the thickness of mudstone in the abandoned river to the thickness of the sand at the corresponding point bar [1, 2].

2.1 Logging acquaintance of abandoned channels

The abandoned channel on a single well is shown that the bottom layer should be equal to the bottom layer of the river sand, and the top layer of the abandoned channel should be lower than the top layer of the river sand (Figure 1). The log curve at the bottom of abrupt abandoned channel is represented as Spontaneous Potential and Natural Gamma curve is box or clock shaped, and the upper near baseline. The log curve at the bottom of the gradual abandoned channel is basically the same as the abrupt abandonment, while the filling part of the upper abandoned channel is jagged, which represented as big amplitude difference of the microelectric level curve at the bottom and small amplitude difference on the top. It reflects on interactive filling of sand and mud.

The method of using electrical relationship to identify abandoned channel is widely used, but it is polysolvable. The lithology, structure and electrical characteristics of the mudstone section in the upper part of the abandoned channel are similar to those of the flood plain, natural levee and breach fan, which makes it difficult to accurately identify the abandoned channel. abandoned channels are mostly isolated meniscus in plane, and there are also curved bands, whose continuity in transverse and longitudinal profiles is limited. Even for the late river channels with continuous curved bands, the longitudinal profiles in parallel river flow direction are not continuous, so identifying abandoned channels requires higher well pattern density. It is necessary to use the methods of profile identification, plane prediction and three-dimensional space combination to identify the abandoned channel comprehensively, but the trend and scale of the abandoned channel cannot be accurately judged based on the logging response of a single well.

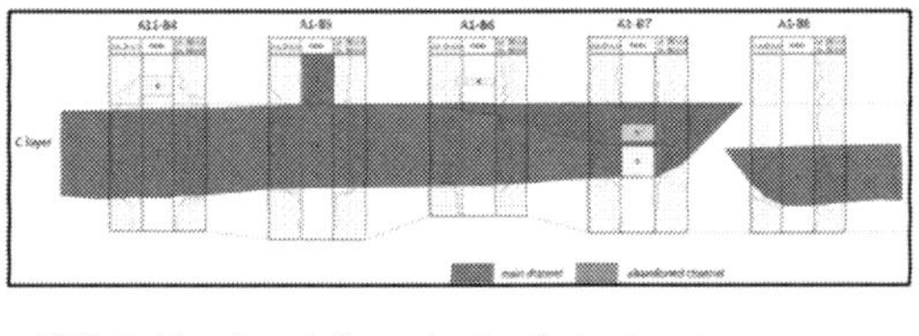

FIG. 1 Abandoned channel microfacies logging response

2.2 Prediction of abandoned river in plane

(1) Sand thickness difference method to identify abandoned channels

The possible distribution locations of abandoned channels can be preliminarily identified through the contour map of sand body thickness. The point dam is relatively thick and in the shape of ring layer, and the abandoned channel adjacent to the river channel is located at the concave bank of the sand body of a single river channel [3, 4]. Because the logging response of abandoned channel is mostly mudstone, the abandoned channel can be roughly identified by the contour map of sand thickness in the data of underground dense well pattern. First, the single stage channel boundary is identified between the point par and the point par (the attribution of the abandoned channel is determined by logging curve or the thin sand bodies between rivers), and the abandoned channel is concave toward the point bar at the channel boundary and the thinning of the channel sand body.

According to the plane morphology and distribution characteristics of abandoned river, The plane combinations of abandoned channel can be divided into two categories: one is the abandoned river formed by single flow, including serial channel type (C type), neck cut type (O type) and breach diversion type (S type); The other type is the abandoned river formed by multi-flow action, such as the abandoned channel of stream diversion type (X type). On the facies zone diagram, the gradually abandoned channel generally presents a discontinuous meniscus shape with small curvature, and the development scale of point bar sand body is small. However, the abrupt abandoned channel is generally independent, whose curvature is relatively large.

(2) Sand top relative depth method to identify abandoned channels

The sand top relative depth method is to predict the plane position of the abandoned channel according to the sand top relative isodepth map on the plane, and then identify the abandoned channel according to its characteristics on the log curve. The relative depth of the sand top refers to the distance between the top of the structure and the sand top of the channel in the small layer. The relative depth of the sand top of the single well of the abandoned channel can be calculated after the single layer top structure is flattened. The relative depth of the sand top of the single well of the abandoned channel is obviously greater than that of the adjacent well, so the position of the abandoned channel can be initially identified by the relative depth map of the sand top [5].

2.3 3D window multi-angle identification of abandoned river

The identification method of two-dimensional plane and section cannot grasp the scale of abandoned channel from a macro perspective, nor can it well control the combination of abandoned river. The multi-angle grid analysis technology in 3D window uses the representation of grid diagram in 3D window to observe and analyze the connected well section to identify abandoned river. By combining the size, direction and shape of the abandoned channel, this method can overcome the defects of two-dimensional identification of the abandoned channel and identify the abandoned river more accurately.

3. Identification of abandoned channels in the study area

The sand body is widely distributed in the plane, and the sediment thickness is also large. The source direction is near northwest, and it is deposited in distributary channel of delta plain. It belongs to meandering river deposit. Most abandoned channels are S-shaped or meniscus shape on the plane.

According to the difference of effective sandstone thickness, sand thickness contour map of C layer in B area is made, as shown in FIG. 2. The contour map of S-shaped abandoned channel on the sand thickness contour map can be obtained by counting the single cycle sand thickness of all Wells, using inter-well interpolation to obtain the contour map, and then delineate the maximum sand thickness to form multiple "beading" local sand body enrichment centers. It fits well with the distribution pattern of point bar in meandering river sediments, and the point bar are regularly distributed along the flow direction. Based on the delineation of local sand thickness centers, the tangential method is adopted to connect several sand-rich centers in turn.

Figure 3 is the relative isodepth map of C layer sand top in B area. The low value in the figure indicates that the relative depth of sand top is small, that is, the thickness of sand body is large, and the relative depth of sand top gradually increases after the transition to high value. Therefore, abandoned channel is most likely to develop in the area with high value. Based on well logging response on the profile and relative equidepth map of sand top, the multi-view grid map is combined in the three-dimensional window, and the abandoned channels that may be missed are identified on the multi-view sedimentary profile according to abandoned channel plane morphology. Finally, the well points of the abandoned channels identified on the plane are reasonably combined according to the sedimentary origin to obtain the distribution of the abandoned channels on the plane.

As shown in Figure 4, well A9-31 has a similar log shape to well A9-32, corresponding to the bottom formations. The thickness of well A9-31 is thinner than that of Wells A9-32 and A9-33 on both sides. It is presumed that the channel boundary of a single phase is between A9-33 and well A9-31. The boundary line to well A9-32 is presumed to be abandoned channel deposit, and the abandoned channel is located on the concave bank of the point point on the plane. If the abandoned channel is on the same side and close to the single-stage channel boundary, the abandoned channel can be continuously combined. Well A9-D42 is in the single-stage channel boundary, but belongs to the off-surface reservoir, and is presumed to be the flood plain facies deposit.

With this method, 3 S-shaped and 7 crescent-shaped abandoned channels were identified in the study area (FIG. 5, FIG. 6). In the sand body of C layer composite river channel in the east of B area, the abandoned channels in the east and west are relatively continuous, and the abandoned channels in the west is characterized by high curvature meandering river, which is presumed to be a abrupt abandoned channel. The central and southern abandoned channels show crescent shape and discontinuous distribution, which is presumed to be gradual abandoned channels. Well A9-F38 is located on the concave bank of a continuous abandoned channel and is a breach fan deposit (FIG. 7).

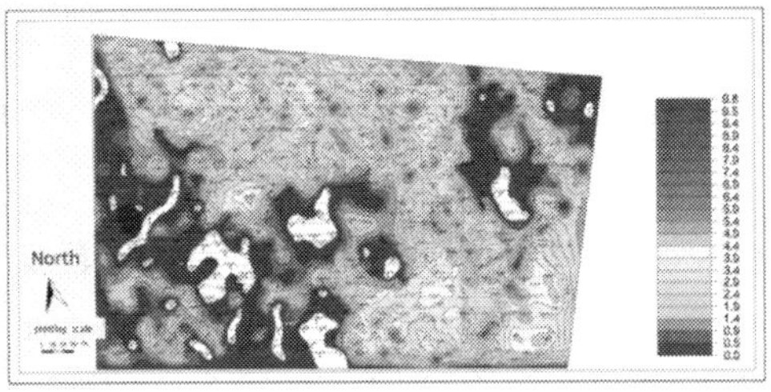

FIG. 2 Effective thickness contour map of C Layer sandstone in B area

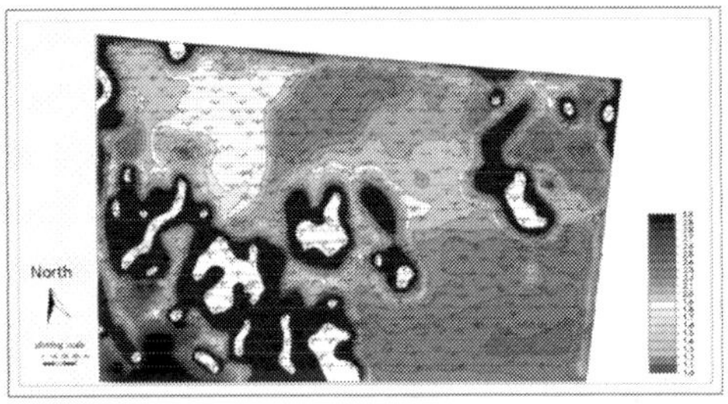

FIG. 3 Relative equal depth map of C layer sand top in B area

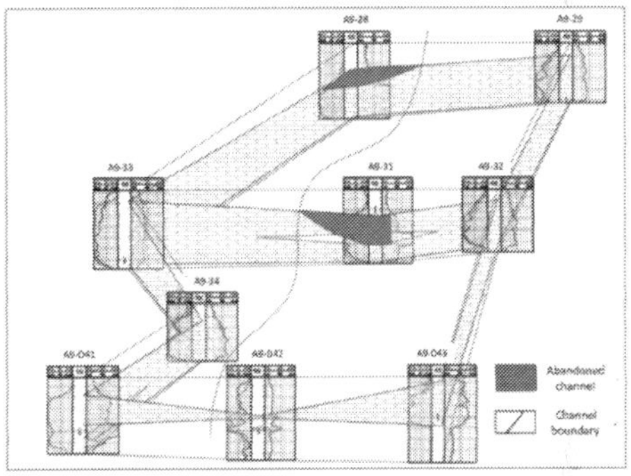

FIG. 4 Three-dimensional grid diagram of C layer in B area

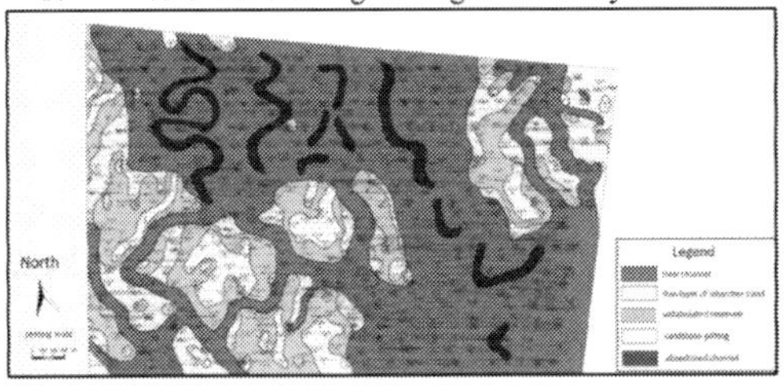

FIG. 5 Sedimentary facies zone of C layer in B area

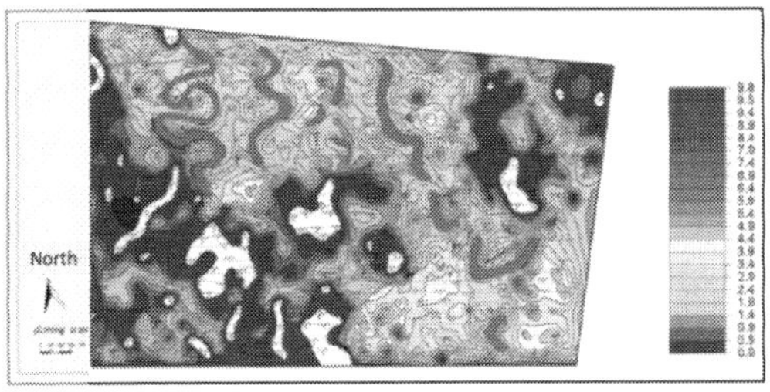

FIG. 6 Abandoned channel distribution in effective thickness contour map of C Layer sandstone in B area

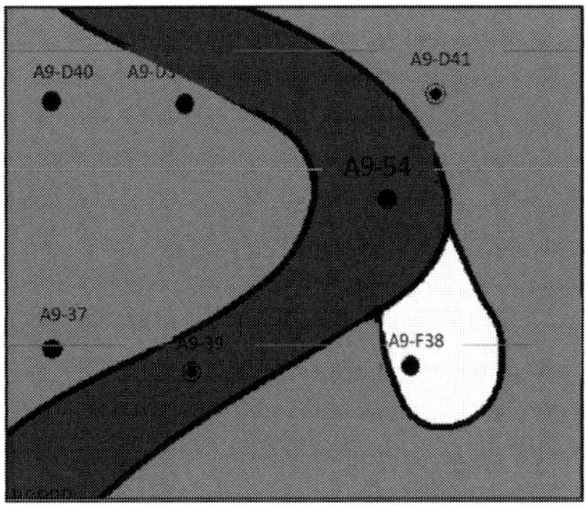

FIG. 7 Well A9-F38 breach fan deposition

4. Conclusions

Under the condition of dense well pattern, the research idea of effectively identifying abandoned channels based on logging data is as follows: guided by the modern meander river sedimentation model, and on the basis of fully analyzing the characteristics of various types of abandoned channels, the method of logging phase identification, effective sandstone thickness method, sand top relative depth method, three-dimensional window grid map and other methods are integrated to cut into and verify each other from multiple angles, and finally identify abandoned channels. Through the application of different methods, three S-type abandoned channels and seven crescent type abandoned channels were identified in the study area, and the western abandoned channels showed the characteristics of continuous and high curvature, so they should be abandoned abruptly. The central and southern abandoned channels show a crescent shape and discontinuous distribution.

References

1. Lei Chen, Jianjun Xu, Shuang Chen, Hui Yang, Linhu Liu. Power Quality Disturbance Signal Denoising Based on Overcomplete Representation. IEEJ TRANSACTIONS ON ELECTRICAL AND ELECTRONIC ENGINEERING. 2022; 17: 544–555
2. Xu Jianjun, Huang Lida, Yan Limei, Yi Na. Insulator Self-Explosion Defect Detection Based on Hierarchical Multi-Task Deep Learning. Transactions of China Electrotechnical Society, 2021, 36(07):1407-1415.
3. Yan Limei, LIU Yongqiang, XU Jianjun, et al. Broken string diagnosis of composite insulator based on Grabcut segmentation and filler area discrimination. Power System Protection and Control, 2021, 49(22):114-119
4. Yi, Q. Wang, L. Yan, et al., A multi-stage game model for the false data injection attack from attacker's perspective. Sustainable Energy Grids & Networks 28 (2021).
5. Na Yi, Jianjun Xu, Limei Yan, Lin Huang. Task Optimization and Scheduling of Distributed Cyber-physical System Based on Improved Ant Colony Algorithm. Future Generation Computer Systems, 109(Aug. 2020), 134-148.
6. Yang Zhao, Jianjun Xu, Jingchun Wu. A New Method for Bad Data Identification of Oilfield Power System Based on Enhanced Gravitational Search-Fuzzy C-Means Algorithm. IEEE Transactions on Industrial Informatics. VOL. 15, NO. 11, NOVEMBER 2019 5963-5970

Study on geological reunderstanding and adjustment potential of development zone

Huiyan Wang

No.5 Oil Production Plant of Daqing Oilfield Co, Daqing 163513, China

Abstract: Development Zone A is a heterogeneous oil field with continental river-delta deposits. The channel sand body is small in scale and is dominated by straight distributary channel sand body. This type of sand body development scale is small, the drilling rate is low, generally only 17.71%, the plane shows a strong heterogeneity, the well pattern control degree is low, the sand body prediction accuracy under the current well pattern condition is low, generally only about 70%. Due to the small scale of sand body development and the development of faults, the contradiction of imperfect injection-production relationship of single sand body is prominent. In 2016, the expansion Wells of Development Zone A were put into operation successively. Based on the re-understanding of the geology of the new drilling area, the adjustment potential of development Zone A was studied, the injection-production relationship of single sand body was clarified, and the adjustment potential was implemented in combination with the results of multidisciplinary reservoir research, so as to achieve the purpose of improving the water drive control degree of the block and improving the injection-production relationship of sand body. After careful analysis, it was preliminatively determined that 19 Wells adjusted for injection and production system were reinjected (6 Wells were reinjected after filling holes). Internal supplementary Wells 19; One oil well was regenerated into a water well, and one water injection well was restored. A total of 41 Wells.

Key words: Heterogeneous oilfield; Fault development; The injection-production relationship.

1. A Development zone basic overview and development status

The northern development zone of T Oilfield (referred to as Development Zone A) was discovered in June 1960 by the first industrial oil flow obtained from well T2. The oil bearing area of Pu1 Formation is 110.02km^2, with geological reserves of 5289.16×10^{4}t and technical recoverable reserves of 1459.85×10^{4}t. The oil layer of Sa 2 Formation developed in the north of T14 row, with an oil bearing area of 19.3km^2 and proven oil reserves of 375.67×10^{4}t. Development Zone A was put into water injection development in 1980, and has successively carried out one infusing adjustment and local supplementary adjustment, development of block T19, expansion development, T14 row to Beisha Group 2 oil reservoir development, and achieved good development results. As of September 2019, the daily fluid volume of the area is 5854t, the daily oil production is 486t, and the comprehensive water cut is 91.69%. The average flow pressure is 3.62MPa, the cumulative oil production is 1187.68×10^{4}t, and the recovery degree is 24.06%.

1.1 The number of water injection Wells is too large, and the degree of water flooding control is low

The oil-bearing area of Development Zone A is 110.02km2. As of December 2016, there are 1,271 oil and water Wells in development Zone A, including 912 production Wells and 359 injection Wells, with a well pattern density of 11.55 / km^2 and a ratio of oil and water Wells of 2.54. A development zone is affected by faults, and the contradiction of imperfect injection and production is prominent. From the table of water drive control degree of different thickness levels (see Table 1), unidirectional and bidirectional connectivity are dominant, and the proportion of connected layers below two directions (including both directions) is 54.7%. The proportion of three-directional sandstone connectivity is only 14.4%.

1.2 The formation pressure level is low and the development effect is poor

From the current pressure level, the formation pressure in the whole area is 9.57MPa, the total pressure difference is -1.61MPa, and the pressure level is low, especially in the well area with imperfect injection and production, the problem of low formation pressure is more prominent.

Since the development of Development Zone A, it has always put the improvement of injection-production relationship and the improvement of ultimate oil recovery in the first place. It has gradually improved the injection-production relationship of single sand body by means of primary infilling adjustment, local infilling and replenishment, and rolling edge expanding drilling, etc. In the high water cut period, multi-disciplinary joint research has quantification of remaining oil distribution. The hard and stable production in high water cut period has been achieved by adjusting the liquid production structure by means of transferring oil well, filling holes and drilling supplementary Wells. However, due to the small scale of sand body development and the development of faults, the contradiction of incomplete injection-production relationship of single sand body is still prominent. From the development effect of development zone A, compared with Xingnan Development Zone, the extraction degree of the same water cut stage is obviously lower, and the development effect is poor. At present, the comprehensive water content has reached 91.69%, and the recovery degree is only 24.06%.

Table 1 Statistical table of water flooding control degree of different thickness levels in development zone A

Effective thickness classification	One way(%)			Two way(%)			Three way(%)			Total(%)		
	Number of layers	sandstone	valid	Number of layers	sandstone	valid	Number of layers	sandstone	valid	Number of layers	sandstone	valid
≥2.0	28.5	32.7	28.3	26.4	30.7	27.1	17.7	20.3	18.6	72.6	83.7	74.0
2.0-1	24.6	29.9	24.5	25.6	32.3	25.9	15.7	21.3	16.0	65.9	83.6	66.3
0.5-1	27.1	32.9	27.2	24.0	30.8	24.2	12.9	18.2	12.8	64.0	81.9	64.1
0-0.5	33.8	42.5	34.0	20.4	25.3	20.1	6.0	6.1	6.2	60.2	74.0	60.3
Extra-surface	37.9	39.4	0.0	20.1	20.6		3.7	3.9	0.0	61.8	63.8	0.0
Total	32.5	34.8	27.1	22.2	28.0	25.5	8.7	14.4	15.5	63.4	77.2	68.1

2. Structural changes and adjustment potential before and after well earthquake combination

From 2011 to 2012, our factory established the structural model of development Zone A based on well vibration. In the following years, the model was updated in time according to the new drilling data. In recent years, new seismic interpretation data have been obtained in an area of about 55km2 in block T30. In order to improve the structural understanding accuracy of this block, the well-earthquake combined structural model was re-established in block T30 in 2016 and integrated into the structural model of large blocks.

2.1 Study of fault variation

First, new faults: 17 new faults, of which 8 are inside the work area, with small fault spacing and many suspected faults. There are 9 peripheral faults with large extension length. Most of the new faults in the working area are at the edge of development Wells. The second is the verification fault: it is mainly the fault with short extension length and small fault distance near the large fault. Verify 4 faults, and verify the existence of small faults without breakpoint control through seismic profiles. The third is the fault morphological change: the fault morphological change is 1, which is mainly a straight extension into a snake extension. The fourth is the extension length change: including the extension length of fault extension and shortening two types. It mainly refers to the fault that the main part of the fault is in good agreement, and the length of the two ends is inconsistent. There are 2 faults in total, and most of the faults are distributed on the edge of the developed well. The fifth is strike change: mainly refers to the fault that the main part of the fault is in good agreement, while the fault tail strike changes greatly. Based on the above analysis, the changes of faults in development Zone A are summarized: 8 faults are newly discovered, 4 faults are cancelled and 5 faults are changed within the comparison boundary before and after 3D seismic interpretation.

2.2 Adjustment potential of fault-changing well area

Due to the small fault spacing, most of them are suspected faults and are located at the edge of the development well, which cannot be verified with the existing data; Due to the fault verification, the originally considered disconnected Wells can form injection and production relationship. For this kind of situation, to analyze the connected relationship between oil and water Wells, and the comprehensive analysis of the oil and water well production condition, the well injection, or combined with the multidisciplinary research results, on the basis of understanding the remaining oil distribution for supplementary well potential, perfect single sand injection-mining relationship, the purpose of mining oil layer residual potential. For example, well area 45-57: the original cognitive structure is a small fault between well

45-57 and well 45-58. After knowing the new structure, the old fault does not exist. The oil Wells in 45-57 well area are relatively concentrated, and the contradiction between imperfect injection and production is prominent. Well 45-57 was designed to inject 4 wells 44-57, 45-56, 45-58 and 46-57.

3. Geological reunderstanding and adjustment potential of new drilling area

With the completion of seismic data interpretation in the edge area of the oilfield and the deployment of evaluation control Wells, it has been confirmed that there are local structural favorable areas on the edge of the developed area of the development zone A. In 2015, based on the implementation of the structure, through the analysis of oil and water distribution, it is believed that the local structure on the edge of the oilfield has the potential for external expansion. At the same time, it is proved by the real drilling data that the local well area on the edge has sporadic expansion potential. In September 2015, the Reservoir Engineering Plan of Puyi Group Outward Expanding Well in the Northern Development Zone of T Oilfield was prepared and completed. A total of 57 outward expanding Wells were designed in the 37-54 well area, 21-8 well area, 2-20 well area and local scattered well area, including 47 production Wells, 10 water injection Wells and 2 old Wells (2-20, 21-8). The next generation of the filling hole is used as a production well, and the oil well is injected 1 (22-45). Drilling began in April 2016 and 56 Wells have been drilled so far. There are 149 oil and water Wells in the old expansion zone, including 117 oil Wells and 32 water Wells. The proportion of local outspread Wells accounted for 27.32% of the total number of all outspread Wells. In order to further understand the characteristics of sand bodies, geological re-understanding of Development Zone A was carried out in 2017.

3.1 Development of newly drilled reservoir

56 local expansion Wells were drilled in Development Zone A, with an average sandstone thickness of 18.2m and effective thickness of 5.8m per well drilled. The thickness of sandstone in the pure oil section is 9.2m and the effective value is 3.8m. The thickness of sandstone in the same interval is 4.9m, and the effective value is 2.0. The thickness of sandstone in the water layer is 4.2m. (See Table 2) From the perspective of oil intervals, well area 21-8 has the largest oil layer thickness and scattered Wells have the smallest drilling thickness. In terms of the development of different thickness classes from new drilling, the off-sheet reservoir and water layer are relatively developed (see Table 3).

Table 2 Average development of newly drilled single well in Development Zone A

well group	Number of Wells	Oil layer		co-layer		water layer		Total	
		Sandstone (Meter)	Valid (Meter)	Sandstone (Meter)	Valid (Meter)	Sandstone (Meter)	Valid (Meter)	Sandstone (Meter)	Valid (Meter)
21-8 well group	12	10.3	5.4	3.5	1.6	5.5	0.0	19.3	7.0
2-20 well group	5	21.9	3.0	2.1	1.0	12.3	0.0	36.2	4.1
37-54 well group	34	7.0	3.6	5.6	2.2	2.4	0.0	15.0	5.9
Scattered well group	5	8.3	2.0	5.8	2.5	5.2	0.0	19.3	4.5
Total	56	9.2	3.8	4.9	2.0	4.2	0.0	18.2	5.8

Table 3 Development of new drilling in different thickness classes in Development Zone A

Effective thickness classification	Oil layer			co-layer			water layer			Total		
	Number of layers (Number)	Sandstone (Meter)	Valid (Meter)	Number of layers (Number)	Sandstone (Meter)	Valid (Meter)	Number of layers (Number)	Sandstone (Meter)	Valid (Meter)	Number of layers (Number)	Sandstone (Meter)	Valid (Meter)
≥2.0	23	69.9	62.6	16	50.4	45.4				39	120.3	108
2.0-1	60	112.3	83.4	25	42.4	33.4				85	154.7	116.8
0.5-1	69	84.2	48	37	46.3	25.6				106	130.5	73.6
0-0.5	70	68.6	19.8	27	23.7	8				97	92.3	27.8
Extra-surface	253	179.3	0	179	109.1	0	175	233.3	0	607	521.7	0
Total	475	514.3	213.8	284	271.9	112.4	175	233.3	0	934	1019.5	326.2

3.2 Geological reunderstanding of the new drilling area

Development Zone A is a heterogeneous oil field with continental river-delta deposits. The channel sand body is small in scale and is dominated by straight distributary channel sand body. Because of the poor continuity of the sand body, in order to better describe the change of the sand body, the method of stratification analysis of the well area is adopted according to the sedimentary facies zone map. Through the comparison of the new and old phase zone maps, the development characteristics of sand bodies are determined, and the continuity, swing position and combination relationship of sand bodies in narrow and small channels are changed after newly drilled supplementary Wells. There are four main types:

Type 1: Newly added channels. There are mainly new narrow and small channels and wide channels. The wide channels are those with a width of more than two well distances, and the narrow and small channels are those with a width of about one well distance. For example, in the Pu-I6 unit in the well area of 20-35, the original river channel was known to flow along the direction of 21-35 and 22-F35, but after the completion of the supplementary Wells, it was found that 7 Wells such as 19-34 and 21-34 were river channels, so the branch rivers of the big river were added. In the expansion block, narrow and small river channels are mainly added.

Type 2: Continuous river becomes discontinuous. For example, in unit Pu-I21 in well area 22-F35, two Wells 22-F35 and 26-F37 in the old phase zone diagram are located in a river channel, and the two Wells are predicted to be a river channel. After the well pattern was encrypted, two new Wells were deployed between the two Wells, one well was drilled off the surface, and the other well was drilled into the pinout zone. Therefore, the predicted channel between the original two channel Wells does not exist. The continuous channel becomes discontinuous.

Type 3: River channel Changes in strike before and after expansion. For example, in the Portuguese I22 unit of well area 37-54, 37-54 in the old phase zone diagram is located in the channel, which is predicted to follow the diagonal direction (no old well). After the completion of the supplementary Wells, the new Wells 38-54, 39-52 and 40-51 were non-channel Wells, while the two Wells 35-S55 and 36-S55 were channel Wells, so the direction of the channel changed.

Type 4: Deregistration of river channels. For example, in Unit Pu-I6 in well area 41-54, 41-54 is an old channel well, and the right channel is a predicted channel. After the completion of supplementary Wells, 36-S53, 37-S53, 38-S53, and 39-54 are located in the channel, but the Wells drilled are all non-channel Wells, so the right channel is written off. And the main river widened.

3.3 New drilling area adjustment potential

After the local outward expansion of development zone A, the development status of the oil reservoir in the new drilling area is analyzed, and the location of the supplementary well is determined or the injection and production system is adjusted according to the sand body. The change types of injection-production relationship in the new drilling area are summarized, and the purpose of perfecting injection-production relationship and tapping remaining potential is realized by adopting effective adjustment measures. For the well areas where injection and production are not perfect due to the absence of injection direction or single injection direction, the well should be reinjected or supplementary Wells should be laid. The single sand body with injection and no production is caused by the perforation of water well but no perforation of oil well. In the external expansion well, the centralized well area and scattered well group are analyzed in detail, and the connectivity relationship of sand body and injection-production relationship are sorted out one by one. Combined with the numerical simulation results, 2 Wells are finally determined to supplement and 7 Wells are injected.

4. Study on adjustment potential and well layout

4.1 New drilling zone adjustment potential

After geological reunderstanding is carried out, the results are applied to multidisciplinary research work to guide the water fitting and remaining oil quantification of multidisciplinary reservoir research, so as to improve the accuracy of water fitting and remaining oil description in A development zone. According to the quantitative remaining oil analysis, the remaining oil in the narrow and straight channel sand of the front phase in the A development area accounts for 57.11% of the total remaining oil of the PuI group, and the remaining oil of the incomplete injection and production type accounts for 24.58% of the remaining oil of the inner front phase, which is the main object to improve the injection-production relationship and tap the remaining potential.

4.2 Formulate adjustment principles

(1)To improve the injection-production relationship in the well area by drilling supplementary Wells or transferring oil Wells in the non-injection direction or in the area with less injection and more production; (2) In the well area with more injection and less production and lack of production well point, the injection-production relationship in the well area should be improved by drilling supplementary oil Wells; (3) In the well area where the injection-production relationship changes greatly due to fault changes after the combination of well earthquake and re-understanding of structure, the injection-production relationship in the well area is improved by drilling and replenishing oil and water Wells or transferring oil Wells; (4) The thickness of sandstone drilled in a single supplementary well is not less than 5.0m, and the effective thickness is not less than 1.0m; (5) In order to prevent casing damage, water injection Wells are not distributed within 100 meters from the fault; (6) Transfer well, not less than 100 meters away from the fault, increase the recoverable reserves of more than 1500t.

4.3 Implementation of adjustment potential

After the preliminary implementation of the well location, the drillable thickness of a single well should be predicted based on the development status of the sand body in the supplementary well area, the utilization status of the single sand body of the well group should be analyzed based on the fine drawing results, the initial water cut and productivity should be predicted based on the results of multidisciplinary research, and the economic benefits of a single well should be evaluated. The well locations that meet the requirements of well layout are designed.

According to the above methods, a total of 19 supplementary Wells (12 production Wells, 7 injection Wells) were finally determined, 1 well was updated into a well, and 19 injection Wells were converted.

5. Conclusions

(1)Deepening geological understanding and perfecting injection-production relationship are the necessary means to improve the development effect of narrow and small channel developed oil fields; (2) Multidisciplinary reservoir study and detailed reservoir description are key techniques to deepen geological understanding and find residual potential; (3) It is an effective way to excavate residual oil in local well areas with imperfect injection-production relationship and improve the development effect of the block through replenishment and transfer.

References

1. W.H.I Mon, Translated by Wang Zhihai et al. Data Warehouse. China Machine Press.2000:1,66
2. Xu Jianjun, Huang Lida, Yan Limei, Yi Na. Insulator Self-Explosion Defect Detection Based on Hierarchical Multi-Task Deep Learning[J]. Transactions of China Electrotechnical Society, 2021,36(07):1407-1415.
3. Limei,LIU Yongqiang,XU Jianjun,et al.Broken string diagnosis of composite insulator based on Grabcut segmentation and filler area discrimination[J].Power System Protection and Control,2021,49(22):114-119
4. Yi, Q. Wang, L. Yan, et al., A multi-stage game model for the false data injection attack from attacker's perspective. Sustainable Energy Grids & Networks 28 (2021).
5. Na Yi,Jianjun Xu,Limei Yan,Lin Huang. Task Optimization and Scheduling of Distributed Cyber-physical System Based on Improved Ant Colony Algorithm. Future Generation Computer Systems, 109(Aug. 2020),134-148.
6. Yang Zhao, Jianjun Xu, Jingchun Wu. A New Method for Bad Data Identification of Oilfield Power System Based on Enhanced Gravitational Search-Fuzzy C-Means Algorithm. IEEE Transactions on Industrial Informatics. VOL. 15, NO. 11, NOVEMBER 2019 5963-5970
7. Jing Han,Xi Wang,LiMei Yan, Aida Dahlak, et al. Modelling the performance of an SOEC by optimization of neural network with MPSO algorithm. International Journal of Hydrogen Energy, Volume 44, Issue 51, 22 October 2019, Pages 27947-27957.https://doi.org/10.1016/j.ijhydene.2019.09.055

https://doi.org/10.1051/e3sconf/202447801013

Development of A Vertical Wire Clamp Mechanism and Wiring Device

Pan Zhou, Song Huang, Keyu Chen, Yongqiang Zhu, Weiyu Zhou, Ren Li

State Grid Yueyang Power Supply Company, Yueyang, Hunan 414000, China

Abstract. With the improvement of social productivity, high-quality power supply is the main requirement for the power industry. To ensure the stability of power supply, power supply companies need to regularly maintain and repair power equipment, during which the use of grounding wire devices is necessary to protect the working process of maintenance personnel. The existing TV, lightning arrester, and circuit isolation switches of 110kV and above grades in substations mostly use the vertical wire connection method to connect with the equipment. During equipment maintenance, there is a risk of virtual connection and dropping when maintenance personnel install grounding wires on such wires, which threatens the safety of the power grid and personnel. To ensure smooth and stable substation equipment maintenance, a vertical wire clamp mechanism and wiring device have been developed, which not only improves the efficiency of maintenance operations and reduces the labor intensity of maintenance personnel but also reduces the work risks during maintenance operations, and is worth promoting in the industry.

Keywords: Maintenance operation; grounding wire; substation equipment; vertical wire.

1. Introduction:

The existing TV, lightning arrester, and circuit isolation switches of 110kV and above grades in substations mostly use the vertical wire connection method to connect with the equipment[1]. Using existing grounding devices on vertical wires often takes a lot of time to connect the grounding wire, and it is still difficult to install[2-4]. Even if it is successfully installed, the contact area between the grounding wire and the vertical wire cannot be firmly secured, making it prone to detachment, which poses a serious risk of electric shock to maintenance personnel[5]. When maintenance personnel loosen the equipment clamp at the bottom of the vertical wire, the bottom of the vertical wire is no longer fixed, and the grounding wire is very easy to drop. At this time, an elevating vehicle or ladder is required to install the grounding wire, which poses a risk of electric shock and can easily cause safety accidents[6].

2. Analysis of the Current Situation

2.1 Demand Analysis

Researchers conducted on-site investigations and field tests in 85 110kV and above substations of a power supply company in a certain city. It was found that 52 of them adopted the vertical wiring method, and the vertical wiring method was present in every non-GIS equipment 220kV substation. Figure 1 shows the vertical wire connected to the lightning arrester.

Figure 1. Lightning arrester with vertical conductor connection mode

The vertical wiring method is widely used in various substations, and the insufficient coordination between the grounding wire and the vertical wire is a common problem, which has a wide market demand and an urgent need for solution.

* Corresponding author: 909617978@qq.com

 https://doi.org/10.1051/e3sconf/202447801013

2.2 Analysis of the Causes

Researchers conducted on-site tests in substations and compared the stability of the connection between the grounding wire and the vertical wire by observing the connection points, testing the probability of the grounding wire falling off after swinging it over a hundred times, and measuring the force required to remove the grounding wire. The results for the vertical wiring method and the horizontal wiring method are shown in Table 1.

Table 1. Stability of matching between equipment and grounding wire

Connection mode	Vertical conductor connection	Horizontal conductor connection
Connection point observation	Contact is oblique, and the connection point is not firm.	Contact is vertical, and the connection is reliable.
Falling rate after swinging 20 times（%）	23%	1%
The Mean force required to remove the grounding wire（N）	610	1120

Researchers compared the conductivity of the connection between the grounding wire and the vertical wire and the horizontal wire by measuring the loop resistance at the connection points. The results for the vertical wiring method and the horizontal wiring method are shown in Table 2:

Table 2. Conductivity of equipment and grounding wire

Connection mode	Vertical conductor connection	Horizontal conductor connection
loop resistance 1（uΩ）	159.5	55.2
loop resistance 2（uΩ）	158.4	56.1
loop resistance 3（uΩ）	163.4	56.2

After three measurements, the results were found to be repeatable, proving that the disturbance tolerance and contact resistance of the vertical wiring method with traditional grounding wire connection are inferior to that of the horizontal wiring method. The specific reason is that when the vertical wiring method is used for grounding, the connection point is not secure due to the diagonal connection, resulting in insufficient clamping force and stability, making it prone to detachment.

2.3 Risk Analysis

According to the investigation, the existing vertical wiring method is widely used in various substations and has difficulties in reliable compatibility with traditional grounding wires. It exhibits insufficient stability and poor conductivity, which affects the grounding reliability of the equipment.

The power supply company in a certain city has more than 30 major repair and renovation projects, over 200 defect elimination plans, and more than 50 emergency repairs in a year. Over 90% of the maintenance work requires the use of grounding wires. If the contact area of the grounding wire cannot be secured or detaches during equipment maintenance, it poses a great electrical safety threat to the maintenance personnel. When the maintenance personnel disconnect the equipment line clamp at the lower end of the vertical wiring, the lower end of the vertical wiring is no longer secured, making it easy for the grounding wire to detach. In such cases, aerial work platforms or elevating ladders are required to install the grounding wire, which also poses risks of electrical shock and falls from height.

Therefore, the development of a vertical wiring clamp mechanism and wiring device will protect maintenance personnel from electrical safety threats and greatly reduce the time for maintenance personnel to install grounding wires on vertical wiring. It will also help reduce safety accidents caused by poor contact of grounding wires in the field.

3. Technical Plan and Working Principle of Vertical Wiring Clamp Mechanism and Wiring Device

3.1 Determination of Major Factors

An analysis of the unreliable grounding factors is shown in Figure 2.

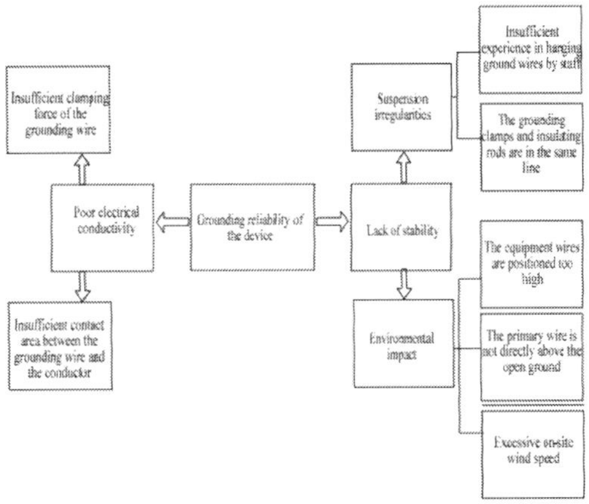

Figure 2. An analysis of the unreliable grounding factors

Based on the identified end factors, the research team has developed a plan for determining the key factors, as shown in Table 3 below.

Table 3. Confirmation schedule of major factors

NO.	Terminal factor	Confirm the content	Confirmation method	Standard	Responsible person
1	Insufficient experience in hanging ground wires by staff	Whether the maintenance level is qualified	Providing technical training and skill assessment for maintenance staff	Whether the qualification rate reaches 100%	A
2	The grounding clamps and insulating rods are in the same line	When grounding the vertical wires, observe on-site whether the contact is good	Record the on-site observations	The contact is good, the wire is completely clamped into the grounding wire clamp, the gap in drop rate is greater than 50%	B
3	The equipment wires are positioned too high	Measure the drop rate of grounding at different heights for vertical conductive grounding	Record and verify the on-site measurements	The gap in drop rate is greater than 50%	C
4	Insufficient clamping force of the grounding wire	Measure the loop resistance after clamping the grounding wire	Record and verify the on-site measurements	The difference is greater than 50 micro-ohms	D
5	The primary wire is not directly above the open ground	Measure the drop rate of the grounding wire on the wall and open ground environment	Record and verify the on-site measurements	The gap in drop rate is greater than 50%	E
6	Excessive on-site wind speed	Measure the drop rate of the grounding wire in different wind speed environments	Record and verify the on-site measurements	The gap in drop rate is greater than 50%	F
7	Insufficient contact area between the grounding wire and the conductor	Increase the area of the grounding wire holding the conductor and measure the DC resistance	Record and verify the on-site measurements	The difference is greater than 50 micro-ohms	G

3.1.1 *Confirmation Process*

The researchers confirmed the cause according to the cause confirmation plan, and the results are shown in Table 4 below:

Table 4.1 Confirmation table of major factors

<table>
<tr><td colspan="2">The staff is inexperienced in hanging the grounding wire</td></tr>
<tr><td>Responsible person</td><td>A</td></tr>
<tr><td>Confirmation method</td><td>Train and test</td></tr>
<tr><td>Confirmation process</td><td>Training and examination were conducted for 15 operation and maintenance personnel, The score is qualified according to 85, and the results shown in the following table are obtained:
<table>
<tr><td>Number of people</td><td>15</td></tr>
<tr><td>Pass rate of theoretical examination （%）</td><td>100</td></tr>
<tr><td>Pass rate of actual operation assessment （%）</td><td>100</td></tr>
<tr><td>Overall situation</td><td>Excellent</td></tr>
</table>
Through the theoretical and practical assessment of 15 professionals, it is found that everyone has qualified level in grounding wire installation and safety measures arrangement, and the grounding wire installation will not be unstable because of their own level.</td></tr>
<tr><td>Conlusion</td><td>Not the main factor.</td></tr>
</table>

Table 4.2 Confirmation table of major factors

<table>
<tr><td colspan="2">The grounding clamps and insulating rods are in the same line</td></tr>
<tr><td>Responsible person</td><td>B</td></tr>
<tr><td>Confirmation method</td><td>Record the on-site observations</td></tr>
<tr><td>Confirmation process</td><td>The researchers used the traditional grounding wire, installed it on the vertical conductor equipment and the horizontal conductor equipment respectively, observed the contact area and cooperation between them, and removed the insulating rod to keep the clamping mouth and manually climbed up to clamp the conductor to re-test and compare:
<table>
<tr><td colspan="3">Test with insulating rod</td></tr>
<tr><td>Connection method</td><td>Vertical wire connection</td><td>Horizontal wire connection</td></tr>
<tr><td>Is the clamp completely clamped?</td><td>NO</td><td>YES</td></tr>
<tr><td>Drop rate after 20 swings (%)</td><td>23%</td><td>0%</td></tr>
<tr><td colspan="3">Take off the insulating rod and test.</td></tr>
<tr><td>Connection method</td><td>Vertical wire connection</td><td>Horizontal wire connection</td></tr>
<tr><td>Is the clamp completely clamped?</td><td>YES</td><td>YES</td></tr>
<tr><td>Drop rate after 20 swings (%)</td><td>0%</td><td>0%</td></tr>
</table>
The results show that when the grounding wire clamp is aligned with the insulating pole in a straight line, it can cause poor connection with the vertical wire equipment, incomplete clamping of the wire into the grounding wire clamp, and reduced reliability.</td></tr>
<tr><td>Conlusion</td><td>The main factor</td></tr>
</table>

Table 4.3 Confirmation table of major factors

<table>
<tr><td colspan="2">The equipment wires are positioned too high</td></tr>
<tr><td>Responsible person</td><td>C</td></tr>
<tr><td>Confirmation method</td><td>Record and verify the on-site measurements</td></tr>
<tr><td>Confirmation process</td><td>Researchers conducted experiments and measurements at different heights to investigate the drop rate of the grounding wire when connected to the vertical wire equipment at different heights:
<table>
<tr><td>Equipment voltage grade</td><td>220kV</td><td>110kV</td></tr>
<tr><td>Drop rate after 20 swings (%)</td><td>23%</td><td>21%</td></tr>
</table>
The results showed that the height had little impact on the reliability of the grounding wire.</td></tr>
<tr><td>Conlusion</td><td>Not the main factor.</td></tr>
</table>

Table 4.4 Confirmation table of major factors

<table>
<tr><td colspan="2">Insufficient clamping force of the grounding clamps</td></tr>
<tr><td>Responsible person</td><td>D</td></tr>
<tr><td>Confirmation method</td><td>Record and verify the on-site measurements</td></tr>
<tr><td>Confirmation process</td><td>During on-site experiments, the researchers measured the resistance between the vertically connected wire equipment and the grounding wire after clamping using additional springs and rupture springs. The results are shown in the following table:
<table>
<tr><td>Connection methods</td><td>Additional spring</td><td>Normal spring</td><td>Breaking the spring makes it weak</td></tr>
<tr><td>Loop resistance 1（uΩ）</td><td>42.2</td><td>60.6</td><td>153.2</td></tr>
<tr><td>Loop resistance 2（uΩ）</td><td>41.5</td><td>58.4</td><td>157.4</td></tr>
<tr><td>Loop resistance 3（uΩ）</td><td>41.7</td><td>63.4</td><td>164.9</td></tr>
</table>
The results showed that the clamping force of the grounding wire clamp had a significant impact on the reliability of the grounding wire.</td></tr>
<tr><td>Conlusion</td><td>The main factor.</td></tr>
</table>

Table 4.5 Confirmation table of major factors

The primary wire is not directly above the open ground			
Responsible person	E		
Confirmation method	Record and verify the on-site measurements		
Confirmation process	The researchers conducted experiments and measurements at wall edges and open spaces to investigate the drop rate of the grounding wire when the primary wire was not directly above the ground.		
	Experiment area	at wall edges	open spaces
	Drop rate after 20 swings (%)	22% (Vertical connection)	23% (Vertical connection)
		0% (Horizontal connection)	0% (Horizontal connection)
	The results showed that the position of the primary wire did not have a significant impact on the reliability of the grounding wire.		
Conlusion	Not the main factor.		

Table 4.6 Confirmation table of major factors

Excessive on-site wind speed				
Responsible person	F			
Confirmation method	Record and verify the on-site measurements			
Confirmation process	The research team members conducted experiments and measurements at different wind force levels to investigate the drop rate of the grounding wire:			
	Wind force levels	1	4	7
	Drop rate after 20 swings (%)	22% (Vertical connection)	25% (Vertical connection)	41% (Vertical connection)
		0% (Horizontal connection)	0% (Horizontal connection)	1% (Horizontal connection)
	The results showed that an increasing wind force increased the risk of the vertical wire disconnecting from the grounding wire but did not affect the horizontal wire connections.			
Conlusion	Not the main factor.			

Table 4.7 Confirmation table of major factors

Insufficient contact area between the grounding wire and the conductor				
Responsible person	G			
Confirmation method	Record and verify the on-site measurements			
Confirmation process	The research team members conducted experiments by installing different numbers of clamps on the same grounding wire to obtain different contact areas with the wire. They measured the corresponding loop resistance:			
	Numbers of clamps	1	2	3
	loop resistance 1 (uΩ)	58.7	52.3	45.1
	loop resistance 2 (uΩ)	60.2	52.4	46.2
	loop resistance 3 (uΩ)	58.6	51.7	45.3
	Increasing the number of clamps did increase the contact area, resulting in a lower loop resistance. However, due to the resistance of the grounding wire and contact resistance, increasing the contact area had little impact on the loop resistance once the clamps were securely connected.			
Conlusion	Not the main factor.			

Based on the above table, there are two main factors that contribute to the low efficiency of protective device replacement, as shown in table 4.8 below.

Table 4.8 Table of main factors

NO.	Main factors
1	The grounding clamps and insulating rods are in the same line
2	Insufficient clamping force of the grounding clamps

3.2 Design Concept

The main factors causing insufficient disturbance tolerance and high contact resistance[7] in the connection between vertical wiring and traditional grounding wires are determined by researchers to be the inclination in the connection between the grounding wire clamp and the insulating rod and the insufficient clamping force of the grounding wire clamp spring[8]. To address these two factors, the researchers propose the following design concepts:

1) To address the issue of alignment between the grounding wire clamp and the insulating rod, the angle between the clamp and the rod can be increased so that the grounding wire can be vertically installed on the vertical wiring equipment, with the contact area between the grounding wire and the vertical wiring equipment accounting for more than 80%.

2) To address the problem of insufficient clamping force of the grounding wire clamp spring, the clamping spring can be replaced with a product with a higher clamping force, ensuring that the circuit resistance between the equipment and the grounding wire is less than 15uΩ when the grounding wire is suspended on the vertical wiring connection device.

3.3 Technical Solution

Researchers have designed a vertical wiring clamp mechanism from the following three aspects, as shown in Figure 3.

1) Angle Design:

To solve the problem of alignment between the grounding wire clamp and the insulating rod, researchers have designed a vertical wiring clamp mechanism, including an installation plate and a clamp block. The installation plate has a wire clamp part, a connecting part for connecting with the supporting rod, and a wiring component for connecting the grounding wire. The wire clamp part has a V-shaped surface facing the clamp block, and the installation plate is L-shaped. The wire clamp part is fixed at one end of the installation plate, forming a U-shaped structure with the installation plate to ensure that the connecting part is parallel to the wire clamp part when clamped. A space for the vertical wiring to enter is formed between the clamp block and the wire clamp part.

2) Clamping Design:

To solve the problem of insufficient clamping force of the grounding wire clamp spring, researchers have decided to place an elastic component between the clamp block rotating device and the installation plate. The elastic component is a torsion spring set on the rotating pin 1. When the clamp block rotates downward, it drives the torsion spring to twist, providing elastic energy storage. After releasing the force that makes the clamp block rotate downward, the clamp block can rotate back under the elastic force of the torsion spring, achieving clamping of the vertical wiring.

3) Overall Design:

Figure 3 shows a vertical wiring clamp mechanism that takes into account both the angle design and the clamping design. During use, the grounding wire is connected to the wiring component, the connecting part is connected to the top of the supporting rod, and the vertical wiring clamp mechanism is supported to the position of the vertical wiring through the supporting rod. Then, the pull rope is pulled downward to enlarge the space between the clamp block and the wire clamp part, facilitating the entry of the vertical wiring. Next, the supporting rod is manipulated to allow the vertical wiring to enter the wire clamp space. Finally, the force on the pull rope is released, causing the clamp block to rotate and tightly hold the vertical wiring between the clamp block and the wire clamp part. Conversely, pulling the pull rope downward and removing the vertical wiring clamp mechanism completes the process.

3.4 Working Principle

The vertical wiring clamp mechanism shown in Figure 3 includes an installation plate 1 and a clamp block 2. The installation plate 1 has a wire clamp part 3, a connecting part 4 for connection with the supporting rod, and a wiring component 5 for connecting the grounding wire. The clamp block 2 is rotatably arranged on the installation plate 1 and is equipped with an elastic element 6 between the clamp block 2 and the installation plate 1. A wire clamp space 7 is formed between the clamp block 2 and the wire clamp part 3 for the insertion of the vertical wiring 8. The clamp block 2 is connected to a pull rope 9 for rotating the clamp block 2.

During usage, the grounding wire is connected to the wiring component 5, the connecting part 4 is connected to the top of the supporting rod, and the vertical wiring clamp mechanism is supported vertically at the position of the vertical wiring 8 using the supporting rod. Then, the pull rope 9 is pulled downward (the length of the pull rope 9 meets the operating requirements and can be designed longer than the supporting rod) to enlarge the wire clamp space 7 between the clamp block 2 and the wire clamp part 3, facilitating the entry of the vertical wiring 8. Next, the supporting rod is manipulated to allow the vertical wiring 8 to enter the wire clamp space 7, and then the force on the pull rope 9 is released. At this time, the clamp block 2 rotates under the action of the elastic element 6, tightly holding the vertical wiring 8 between the clamp block 2 and the wire clamp part 3. On the contrary, pulling the pull rope 9 downward, and then removing the vertical wiring clamp mechanism completes the process. This vertical wiring clamp mechanism is suitable for grounding the vertical wiring 8, with simple operation, good stability after connection, less chance of falling off, and high safety.

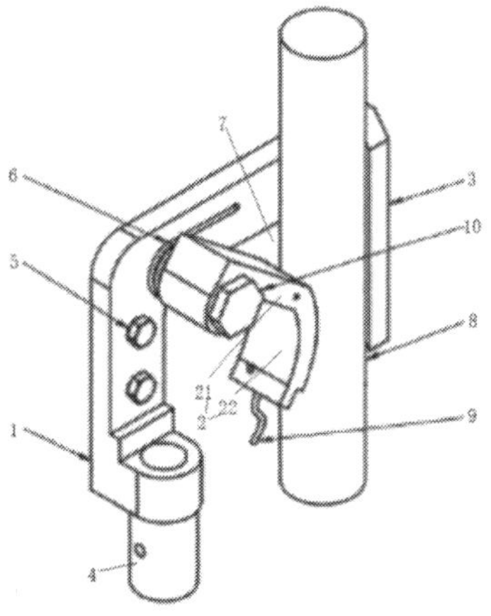

Fig. 3 Structure diagram of vertical conductor clamping mechanism

4. On-site Verification

The clamp mechanism and wiring device were used at a 220kV substation during the complete shutdown and maintenance of the 220kV equipment, as shown in Figure 4. The grounding wire did not fall off during the maintenance process, and the maintenance work was successfully completed. The connection was stable and the contact resistance was low. The results of the field verification are shown in Tables 5 and 6.

Fig. 4 Site diagram of a 220kV substation product.

Table 5. Stability of matching between equipment and grounding wire

Connection method	Vertical wires connection
Observation of connection point	Contact is vertical, and the connection point is not firm.
Drop rate of left and right swing (%)	0%
Proportion of contact cross-sectional area (%)	100%

Table 6. Conductivity of equipment and grounding wire

Connection method	Vertical wires connection
Loop resistance 1 (uΩ)	42.1
Loop resistance 2 (uΩ)	41.8
Loop resistance 3 (uΩ)	41.8

After adopting this new technology, it not only reduces the workload of maintenance personnel, but also greatly improves the reliability and safety level of grounding during maintenance, ensuring the personal safety of the maintenance personnel.

5. Conclusion

The new type of clamping mechanism and wiring device designed for grounding connection of vertical conductors ensures reliable and secure installation, avoiding incidents caused by loose grounding wires [9]. More importantly, it guarantees the personal safety of maintenance personnel. The operation and use of this equipment are simple and convenient, eliminating the need for extensive learning costs associated with tool replacement. It improves the efficiency and safety of substation maintenance operations [10] and has high promotional value.

Acknowledgements

This paper is supported by the Science and Technology Project of State Grid Hunan Electric Power Co., LTD. (Project number: 5216A8220022)

6. References

1. Liang, Liuming, Xie, Lingdong, Zhao, Luzhen, et al. (2016). Development and Application of 220kV Vertical-wire Grounding Lead. Journal of Shanghai Electric Technology, 9(03), 21-23.
2. SHAO Jiankang(2023). Analysis of Fault and Improvement Measures of Substation Grounding Device. Electronic Technology, 52(02), 110-111.
3. Xu, Hui, Lan, Xingwei, Dong, Jiejun, et al. (2021). A Rapid Install-Uninstall Ground Wire Device.

Scientific and Technological Innovation, (01), 163-164.

4. SONG Hengdong;PAN Lingmin;HAN Xuechun; WU wei. (2021). Development and Application of Intelligent Grounding Line. Electric Safety Technology, 23(03), 68-71.
5. Zhang, X. (2023). Reflections on an Accident Caused by Personal Electric Shock. Rural Electrician, 31(05), 56-57.
6. LI Jianpeng;ZHAO Jining;ZOU Yuan;LIU Xiaofei; WANG Xu (2022). Multiple Correspondence Analysis of Influencing Factors of Power Safety Accidents and Optimization of Control Measures. Hebei Electric Power Technology, 41(05), 75-79.
7. Wang, Y. (2022). Study on Contact Resistance Characteristics of Grounding Down Conductor in Substation. North China Electric Power University (Master's thesis).
8. Guo, Pingjun, Wang, Wei. (2013). The device for quickly adjusting clamp force given by belleville spring set. Manufacturing Technology and Machine Tools, (08), 117+121.
9. He, Weidong. (2021). Violations and Non Standard Practices Are the Enemies of Accidents. China Power Enterprise Management, (06), 13.
10. Zhang, Ping. (2019). Analysis of Key Technologies in Maintenance of 110kV and Below Substation Equipment. Telecom World, 26(06), 183-184.

Research on Technical Difficulties and Countermeasures of Well-Seismic Combination Unit-Level Tectonic Modeling

Xiaofang You

No.2 Oil Production Plant, Daqing Oilfield Co., Ltd., Daqing, China

Abstract: Compared with the previous reservoir modeling, the well-seismic combination sedimentary unit level structural modeling is faced with problems such as large amount of data, difficulty in data loading, a large number of inclined wells, causing serious structural distortion, and the local morphology of the fault needs to be accurate to the unit level. In view of the above difficulties, this paper systematically introduces the corresponding countermeasures to solve the corresponding problems. The built model can characterize the unit level structural characteristics with high accuracy, and can guide the fine adjustment and exploration of the sedimentary unit level scale.

Keywords: Structural modeling, sedimentary unit, well-seismic combination.

1. The raising of the question

The study area has entered the ultra-high water cut stage, and the remaining oil is scattered, and the fault edge is the favorable part for remaining oil excavation [1-3]. In order to fully tap the remaining oil at the fault edge and clarify the injection-production relationship at the sedimentary unit level, the well-seismic combined with unit-level structural modeling research has been carried out since 2018.

Different from reservoir group-level modeling, well-seismic integrated sedimentary unit-level modeling is faced with many problems, such as many wells, many layers, large amount of calculation, the scale of fault feature needs to be raised to the unit level, and high structural accuracy requirements. In order to ensure the quality of the model and meet the development needs, the research on the technical difficulties and countermeasures of unit-level modeling is carried out.

2. Technical difficulties and solutions

Compared with reservoir modeling, sedimentary unit-level modeling has four major technical difficulties (Table 1).

Table1 Modeling Technical Difficulties, Countermeasures and Objectives

Number	Technical difficulties	Countermeasures	Targets
1	Inaccurate inclined well data leads to severe structural deformation	Azimuth correction Trend control	Restore the reasonable structural shape of inclined well
2	Large data volume Difficulty loading	Batch loading Dilution treatment	Establish the overall work area
3	The accuracy of fault characterization requires Promote to unit level	Well seismic fault point matching Unit boundary control	Accurate description of the fault Three-dimensional distribution shape
4	Multiple and thin units Adjacent layers are easily intersected	Stepwise constraint Thickness control	Trend of layer model Reasonable and non intersecting

(1) Inaccurate inclined well data leads to severe structural deformation

In practice work, the stratification point of inclined wells often differs greatly from the stratification trend of surrounding wells, which causes the distortion of the structural surface around inclined wells. It may be affected by the data acquisition time, instrument difference, data processing and other factors. Therefore, it is necessary to correct the well deviation data.

The first is azimuth correction, which mainly aims to restore the reasonable position of the plane of the stratification point of the inclined well.

During the drilling process, measuring instruments measure the depth, inclination, and azimuth of each measuring point in the well trajectory. The measurement instruments of azimuth are mostly fluxgate and gyroscope[4-5]. The measurement of well inclination azimuth is usually carried out by magnetic measurement instruments, and the measured azimuth is based on the magnetic north position.

When using non-magnetic measuring instruments such as gyroscopes, the measured azimuth is based on the true north azimuth. However, because the orbit design and trajectory calculation of the directional well are both based on the Gaussian projection coordinate system, which is based on the grid north position. So it is necessary to convert the measured magnetic north as the reference for the azimuth angle of the well deviation into the azimuth angle of the well deviation based on the grid north position. This work is called "azimuth correction" and is referred to as "azimuth reference frame conversion" abroad.

Through the establishment of the model, it is shown that when the local dip angle exceeds 5°, for the directional well with a horizontal displacement of 400 m, if the magnetic azimuth angle correction is not considered, the well trajectory deviation in the X direction is 46.6m, and the deviation in the Y direction is 64.75m [6]. Therefore, it is necessary to carry out the correction work. According to the actual working situation, the correction angle in the research area is generally 10.33°.

The second is to correct the inclined well and restore the reasonable form of the structure.

The structural trend surface method is mainly adopted for correction, that is, the correction amount of inclined well coordinates is determined by analyzing the difference between distorted structural surface and reasonable structural trend surface, so as to further modify the inclined well trajectory (Fig.1, Fig.2).

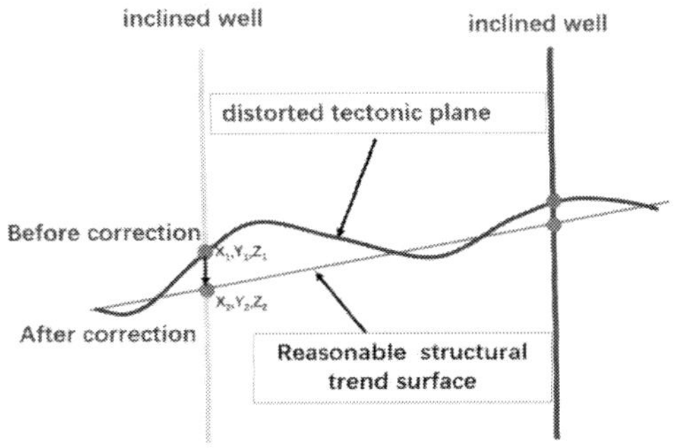

Fig. 1:Schematic diagram of correction principle

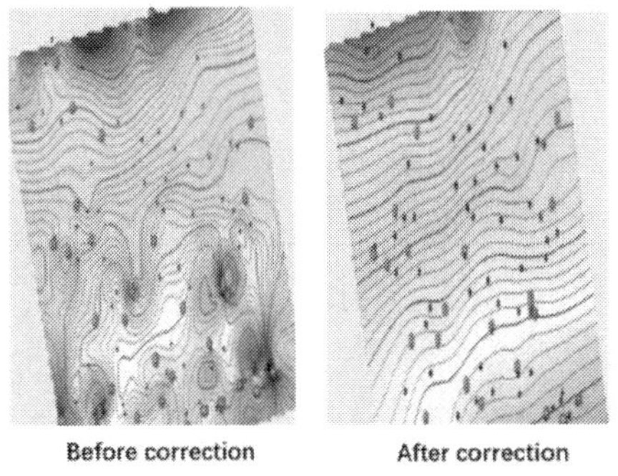

Fig. 2: Structure comparison diagram before and after correction

(2) Large data volume, difficult to load

In the study area, about 102,375 files were used in modeling, and the total number of record points exceeded 100 million, which is a hundred times of the data volume of reservoir group levelmodel. Therefore, we adopted the method of batch loading and thinning processing to reduce the pressure of large data volume and establish a unified overall work area.

One is to break it into pieces and load it in batches.

There are 17,708 wells of all types in the work area, which are loaded in 9 batches, improving the efficiency of the software operation.

The second is to conduct thinning processing on the well deviation data.

The interval of well deviation data measurement is 0.05m, and the record points of each well reach more than 20,000, which means that the record points of more than 17,000 wells reach the level of hundreds of millions, This is a great test for software and hardware. In view of the situation of excessive data volume and difficult loading, under the premise of ensuring accuracy, data extraction is processed, and the original 0.05m interval is extraction to 5m interval, which not only ensures data accuracy, but also ensures the running speed.

(3) The fault characterization accuracy should be improved to the unit level

Compared with reservoir group-level fault modeling, unit-level fault modeling requires more detailed depiction of local fault morphology. In the past, reservoir group-levelfault modeling mainly controlled the overall morphology of faults, focusing on matching breakpoints near the top surface of the oil layer group and those with larger fault spacing with the cross-section. However, it did not achieve accurate depiction of the internal morphology of faults in the oil layer group. Therefore, the trend surface display of reservoir group-level fault morphology can meet the needs. However, for the sedimentary unit-level fault morphology, the main fault point and the accessory fault point must be clearly defined, and the main fault point and the fault must be strictly matched. On this basis, it is further considered whether the unit boundary points on both sides of the fault belong to the correct fault hanging wall or footwall, and the section position is adjusted according to the structural drop between well points. Only by meeting the requirements of strict matching of the main fault point and the section and the

 https://doi.org/10.1051/e3sconf/202447801014

unit-level boundary point belonging to the correct fault plate can the section morphology be accurately depicted to the unit level (Fig. 3,Fig. 4)

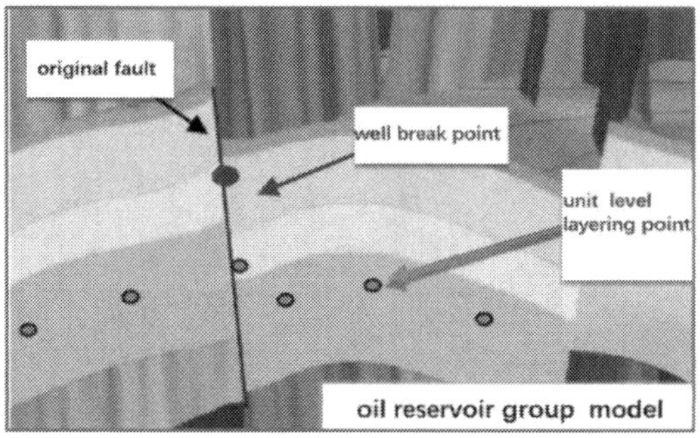

Fig. 3: Location of stratification points and layer morphology before constraint of main break points

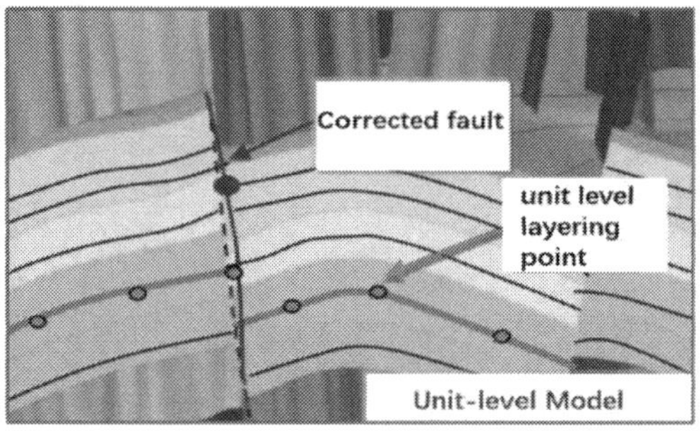

Fig. 4: Location of stratification points and layer morphology after constraint of main break points

(4) Multiple and thin units, with adjacent layers easily intersecting

There are many subdivision layers in the study area, and 195 sedimentary units can be subdivided vertically, the thinnest sedimentary unit measured thickness is less than 1m. When establishing the unit level model, adjacent layers without well control between wells are very easy to cross each other. In view of this difficulty, on the basis of well vibration combined with simulated tectonic trend constraints on the top surface of the reservoir group, the unit level modeling is established by using the method of stepwise constraint and thickness control.

3. Application Case Analysis

The establishment of the sedimentary unit-level structural model further improves the refinement of reservoir description, which can effectively guide the refined adjustment of the oilfield to tap the potential.

1. The integrated application of unit structure map and facies zone map has been realized, which can effectively guide the adjustment of injection-production relationship on both sides of the fault;compared with the reservoir group-level faults, the understanding of the injection-production connectivity of 445 injection-production well groups near 59 faults has been changed, and 49 high-efficiency wells in the fault zone have been further realized (Table 2).

Table 2 Statistical Table of Potential for Supplementary Drilling in Fault Areas

project		Number of wells		
		Oil well	Water well	subtotal
water drive	vertical well	15	14	29
	high angle well	8		8
	subtotal	23	14	37
Three mining and supplementary drilling		10	2	12
total		33	16	49

2. Quantifying the distance between logging and fault unit level, further releasing the well distribution potential of large displacement directional wells and complex structural areas;

(1) In the early stage, a total of 27 highly inclined directional wells were implemented to tap potential, with a daily oil production of 9.2t per well in the initial stage. By accurately characterizing the spatial distribution of faults, we have expanded the potential of 8 wells in the already implemented areas (Fig. 5).

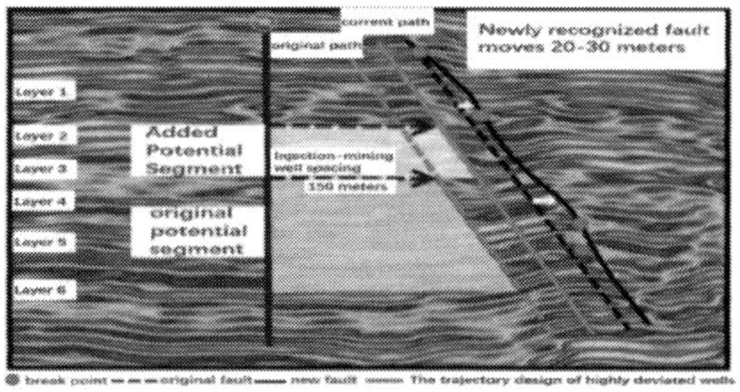

Fig. 5 Principal breakpoint constraint section curvature

(2) Re-recognize the potential of fault block area, and expand the potential of 41 supplementary drilling wells (25 oil wells and 16 water wells) (Fig. 6).

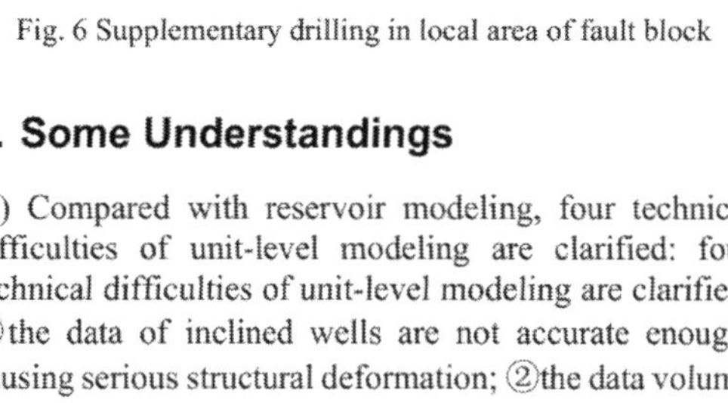

Fig. 6 Supplementary drilling in local area of fault block

4. Some Understandings

(1) Compared with reservoir modeling, four technical difficulties of unit-level modeling are clarified: four technical difficulties of unit-level modeling are clarified: ①the data of inclined wells are not accurate enough, causing serious structural deformation; ②the data volume

is large, and it is difficult to load; ③ the fault characterization accuracy needs to be improved to the unit level; ④the number of units is large and thin, and adjacent units are easy to be intersected;

(2) In view of the technical difficulties, the corresponding countermeasures are formulated. Through azimuth correction, data sampling, batch loading, main fault point constraint surface characterization of fault morphology and step by step constraint, thickness control and other methods, the research goal of establishing the unit-level structural model is realized;

(3) The establishment of the sedimentary unit-level model realizes the integration of two maps, clarifies the unit-level injection and production relationship, quantifies the distance between logging and fault unit level, deepens the understanding of the residual oil potential near the fault, and releases the well distribution potential.

References

1. Gan Lideng, Dai Xiaofeng, Zhang Xin, et al. Key Technologies for Seismic Reservoir Description in High Water Cut Oilfield [J]. Petroleum Exploration and Development, 2012, 39 (3): 365-377
2. Zhou Liqing, Lei Yixin. Formation conditions and exploration direction of large and medium-sized oil and gas fields in the fault depression strata of the Songliao Basin [J]. Petroleum and Natural Gas Geology, 2006, 27 (6): 820-826
3. Xu Hong. The control effect of the Xingbei fault on the formation of reservoirs in the uplift area outside the source of the Fuyu oil layer [J]. Daqing Petroleum Geology and Development, 2013, 32 (6): 28-31
4. Lin Yanbing, Yan Jizeng. A correction method for wellbore azimuth angle based on geomagnetic parameters and meridian convergence angle calculation [J]. Exploration Engineering (Geotechnical Drilling Engineering) 2012, 39 (12): 1-4
5. Wang Jiexian, Li Haojun. Formula derivation and calculation of magnetic declination and tilt angle [J]. Geodesy and Geodynamics, 2009, 29 (3): 1-4
6. Wang Min. Precision Analysis of Geological Models Based on Magnetic Azimuth Correction [J]. Drilling Technology, 2017, 40 (3): 1-3

A New Perspective of Trees Competition Network Research

Shaohua Wang [1], Ting Yang [1], shengxiang Ouyang [2]

[1] University of Arts and Science, Changde 415000, China.
[2] Changde Forestry Science Research Institute, Changde 415000, China.

Abstract: Forest competition is a key factor driving forest dynamics and stand structure, and is one of the core issues in ecological research. Competition is a complex ecological process with temporal changes, which is interrelated and affected with forest growth, stand structure dynamics and environmental changes. Although there have been many researches on competition, tree growth, stand structure and climate, the interactions among trees have been studied by static competition index measurement, which usually lacks temporal variation related to natural forest development and environmental conditions, and the response mechanism and related theory of competition to stand development and environmental fluctuation have been studied insufficiently.(2) Single tree competition index model;(3) Interaction between tree competition and tree growth, stand structure and environment;(4) Tree competition based on complex network. Finally, according to the complex interaction of tree competition, the research strategy of tree competition network structure with "single tree" as node and "single tree competition relationship" as edge is discussed, in order to provide reference for the study of tree interaction and forest ecosystem dynamics.

Keywords: tree competition; competition network; stand structure; environmental response.

1. Introduction

Forest plays an important role in the process of human history and civilization, and plays an extremely important and irreplaceable role in maintaining biodiversity, protecting ecological environment, mitigating natural disasters, regulating global carbon balance and biogeochemical cycle. Tree individuals are an important part of forest ecosystem, and their growth is affected by many external factors besides their own genetic factors. Competition and coexistence among tree individuals is always a core problem in ecological research. Community structure formation, productivity formation, system stability and community species diversity maintenance are closely related to this problem. Competition among plants refers to the interaction between two or more plants competing for the same environmental resources and energy. Competition directly causes differences in tree growth and shape, and is a key factor driving forest growth and stand structure dynamics.

Forest structure reflects individual trees (structural elements) and their attributes Forest structure is the basic unit of forest structure, a comprehensive reflection of forest development process, an important factor in forest management and analysis, and a specific object of forest management activities[1]. As a component of stand structure, microenvironment of individual tree affects the future development of individual tree. Therefore, it is an important technical means of structured forest management to quantify microenvironment of individual tree based on adjacent tree relationship and analyze and study microcharacteristics of stand spatial structure. In the forest complex network structure composed of "individual tree-adjacent tree relationship", tree competition is the main driving force of forest community dynamics, affecting tree growth, stand productivity and stand structure. Some scholars often use interspecific and intraspecific competition index to understand tree interaction[2-4]. Meanwhile, tree growth dynamics, stand structure and external environment also have corresponding influence and restriction on tree competition effect.

In recent years, forest competition is mainly studied by static competition index measurement, which is usually used to analyze and evaluate the competition relationship of trees. This measurement usually lacks temporal changes related to natural forest development and environmental conditions, and the research on forest competitiveness is still mainly concentrated on the construction of index system based on adjacent tree relationship, staying at the level of static rating. Therefore, it is urgent to establish a dynamic theoretical framework of forest competitiveness according to the structural characteristics of natural forest, which is the main body of forest ecosystem, so as to provide new ideas and perspectives for the study of forest competition. This paper reviews the characteristics of the relationship between tree competition and stand spatial structure and its related theoretical studies, mainly including the following aspects: (1) parameters of stand spatial structure;(2) competition index model of individual tree;(3) interaction response relationship between tree competition and tree growth, stand structure and environment;(4) tree competition based on complex network. Finally, the paper discusses the research strategy of tree network structure with "individual tree" as node and "competition relationship between individual trees" as edge, aiming at the complex interaction of tree competition, so as to provide reference for the study of tree interaction and forest ecosystem dynamics.

2. Characteristics and theories of the relationship between forest competition and stand spatial structure

Forest competition is a complex interaction, which directly causes differences in forest growth and characters, and is a key factor driving forest growth and stand structure dynamics, while forest growth dynamics, stand structure and environmental changes also have corresponding effects and constraints on forest competition effects. Network-based analysis is becoming one of the main trends in the study of interactions between biological communities and individuals. However, it has been neglected in the study of vegetation competition.

2.1 Stand spatial structure parameter

Stand microenvironment can reflect the size, distribution, species and extrusion of adjacent trees around the forest, reflecting the microscopic characteristics of the spatial structure of the forest, so scientific and accurate quantification of the microenvironment of individual trees is very important in structured forest management. Hui Gangying proposed a method based on the relationship between adjacent trees to accurately quantify the microenvironment of trees, In structured forest management, angular scale is used to adjust the spatial distribution pattern of forest stand in order to increase the randomness of trees; in order to increase the diversity of tree species, the isolation degree of trees is adjusted by mixing degree; in order to enhance the competitiveness of target trees, the competition relationship of tree species is adjusted by size ratio, and in order to increase the nutrient space of target trees, the crowding degree of trees is adjusted by density.

2.2 Single tree competition index model

Competition among trees is an important part of forest spatial structure [5]. Competition degree among trees is generally expressed by competition index. Competition directly causes differences in tree growth and shape, and competition index quantitatively describes competition among trees, formally reflects the relationship between individual growth and living space of trees, and essentially reflects the demand of trees for environmental resources and the competitive pressure they bear when fighting for environmental resources. Since 1960s, foreign scholars have put forward competition indexes related to distance and unrelated to distance, among which Hegyi competition index is the most widely used. On this basis, Chinese scholars have put forward some improved competition indexes. However, most of these competition indexes only consider unilateral or bilateral factors.

2.3 Interaction among competition, growth, stand structure and environment

Competition is a key factor driving forest dynamics and forest structure during stand development. Competitive indices calculated from competitive trees within a certain range can simulate the growth process of target trees at different time intervals. At the same time, continuous growth of trees regulates stand development, stand productivity and interactions within tree populations.

(1) Competition index is the basis of establishing individual tree growth model, and its quality directly affects the effect of individual tree growth model. Bhandari (2021) compared the validity of different competition indices in growth prediction models, and the results were consistent with those of other scholars. During these studies, scholars believed that changes in inter-tree distance and individual plant size led to changes in competition intensity and its impact on growth, which may be one of the reasons why competition indices had a significant impact on growth prediction[6].

(2) In the aspect of competition and stand structure, Chinese scholars analyzed the correlation between competition and stand density, DBH structure, tree height and crown width. Forrester found that interactions between tree species did not occur or were weak in low-density stands, while interactions between tree species occurred more frequently as stand density increased. Competing understory vegetation could also exert control over tree growth at early stages of stand development. Scholars found that competitive effects increased faster in dense stands and were exacerbated by lower water availability[7].

(3) In terms of competition and environmental factors, Roux[8] believes that climate change may affect individuals interacting with species by studying changes in vegetation interaction along two environmental gradients. Temporal fluctuations in the environment can change interactions between adjacent tree individuals[9], thus affecting the competitiveness of trees[10]. Understanding how competition evolves over time is therefore crucial to assessing forest dynamics under changing environmental conditions. Some scholars suggest that competition may be negatively correlated with climatic conditions affecting tree growth[11]. Other studies have found that tree growth and competition effects are highly dependent on local climatic conditions[12].

Therefore, competition effect of trees changes dynamically with stand growth, and competition pressure between target trees and competing trees may not be constant with time due to environmental changes. Although competition effect of adjacent trees on stand dynamics has been paid more and more attention, the response of competition to environmental fluctuations and stand growth has not been fully explored.

2.4 Competition of Trees Based on Complex Networks

In recent years, network-based analysis is becoming one of the main trends in the study of interactions between biological communities and individuals[13]. Complex network theory has become a new perspective and new idea to reveal various ecological processes, but this method has been neglected in the study of plant competition for many years. Nakagawa[14] used complex network analysis to reveal some basic characteristics of individual competition in the same age stand of Abies curioides; The role of individual nodes in networks (and their removal) has been examined in studies of environmental dependence and plant interactions[15]. Vazquez showed that the intensity of interspecies interactions has a significant impact on the structure and dynamics of ecosystems. Mongus proposed a new approach based on complex networks, which enables higher precision prediction of future growth of individual trees by conceptually incorporating complex network theory into a structured representation of tree competition. Comprehensive analysis shows that tree competition is a key factor driving forest dynamics and stand structure, and is one of the core problems in ecological research. Competition is a complex ecological process with temporal changes, which is interrelated and influenced by tree growth, stand structure dynamics and environmental changes. Although studies on competition, tree growth, stand structure and climate have been carried out, the interactions between trees have been studied through static competition measures, which usually lack temporal changes related to natural forest development and environmental conditions, and the response mechanisms and related theories of competition to stand development and environmental fluctuations are insufficient.

3. Research prospects

3.1 Construction of Structured Forest Competition Network

The driving forces for the formation of structured stand networks were analyzed as shown in Figure 1. According to the spatial structure characteristics of the stand, a structured stand network was constructed by taking individual trees as nodes and "competition relations between individual trees" as edges connecting nodes. The microstructure of the structured stand competition network was studied from three aspects: individual state of trees (nodes), competition relations among trees (relations) and environment (resources) in which trees were located. The structural characteristics and linkage mechanism of structured stand network were revealed by analyzing the formation mechanism of structured stand network.

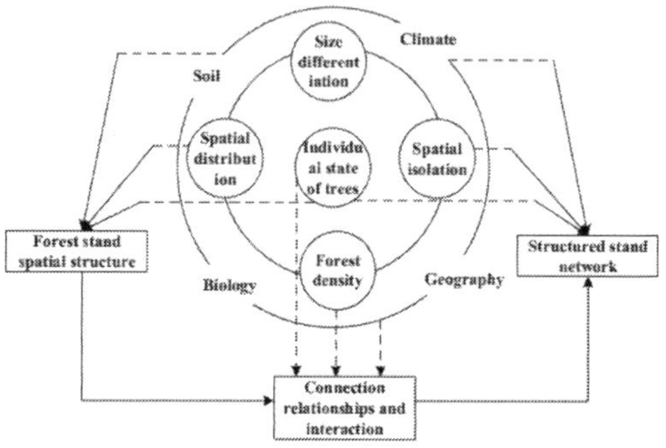

Fig.1 The driving force behind the formation of structured forest stand networks

3.2 Construction of Forest Competitiveness Model Based on Dynamics

The paper analyzes the inherent mechanism and model characteristics of maintaining the structure characteristics of stand competition network, and quantitatively analyzes the spatial structure of stand competition network based on the node "point elements" and the relationship "edge elements" of connecting nodes, as shown in Figure 2. The nodes in the network represent the characteristics and functions of individual trees, and the edges represent the competition relations and intensities of adjacent trees. The state of the whole network is driven by a set of nonlinear parametric equations. By constructing a parametric function fw such that $x_n = f_w(l_n, l_{co[n]}, x_{ne[n]}, l_{ne[n]})$, where $l_n, l_{co[n]}, x_{ne[n]}, l_{ne[n]}$ represent the characteristics of the current node, the edges of the node, the characteristics and functions of the adjacent nodes, respectively. In this undirected graph, the edges of the nodes are represented by adjacency symmetric matrices, and the relationships between all nodes are represented by adjacency matrices. At the same time, the concept of competition field of structured stand network is put forward, that is, the field formed by the definite force (attraction or repulsion) produced by single tree in stand network to adjacent trees in different positions in space. The competition field is divided into scalar competition field and vector competition field. The core elements and structural characteristics of competition field of structured stand network are analyzed and discussed. The static elements and dynamic elements in competition field of network are analyzed and discussed.

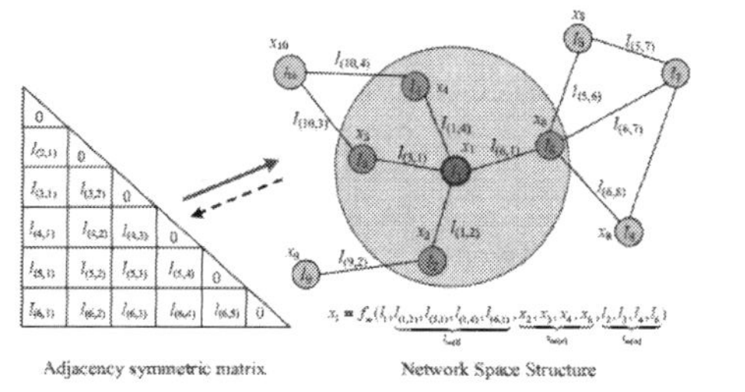

Fig.2 The spatial structure and adjacency matrix of the "node edge" network in forest stands

3.3 Environmental Dynamic Response of Structured Tree Competition Network

The inherent mechanism and characteristics of structured tree competition network responding to environmental dynamics were analyzed and discussed. By collecting and pretreating the standing tree core samples, measuring the annual ring width and bark thickness of the samples, analyzing the radial growth and its annual variation in combination with the stand spatial characteristic indexes obtained from each tree scale of the sample plot, reconstructing the stand spatial structure in different years, analyzing and discussing the response and variation characteristics of the reconstructed stand spatial structure with time and environmental fluctuation in combination with the environmental factor indexes of the corresponding years, By analyzing and discussing the static and dynamic elements in multi-layer temporal competitive network, the fluctuation characteristics of forest competitive network model with multi-layer temporal network structure with time and environment were studied, which provided reference for the study of competition response to environmental fluctuation and forest growth.

To sum up, this paper looks forward to the construction of structured forest competition network and the research of forest competitiveness model based on dynamics theory. According to the microstructure characteristics and link mechanism of structured forest competition network, it is expected to reveal the mode and characteristics of structured forest competition network. At the same time, based on the data of tree core samples, the spatial structure of stand was reconstructed, and the environmental factors were combined to reveal the dynamic response mode and characteristics of multi-layer time-series stand network competition field to the environment, which provided a new idea and perspective for the study of tree competition, and had important significance for further developing the theory and application research of tree growth prediction, stand structure adjustment and biodiversity protection.

Acknowledgments

Hunan University of Arts and Sciences Doctoral Research Initiation Project (20BSQD04): Hunan Provincial Natural Science Foundation (2023JJ60168)

References

1. O'Hara K L, Latham P A, Hessburg P, et al. A structural classification for inland northwest forest vegetation[J]. Western Journal of Applied Forestry. 11: 97-102, 1996.
2. Versace S, Vittorio Garfi, Dalponte M, et al.Species interactions in pure and mixed-species stands of silver fir and European beech in Mediterranean mountains[J].iForest - Biogeosciences and Forestry, 2021, 14(1):1-11.
3. Brunner A, Forrester D I. Tree species mixture effects on stem growth vary with stand density – An analysis based on individual tree responses[J]. Forest Ecology and Management,2020, 473:118334.
4. Ja A, Rpb C, Mdrb C, et al. Species stratification and weather conditions drive tree growth in Scots pine and Norway spruce mixed stands along Europe[J]. Forest Ecology and Management, 2021, 481:118697.
5. Luo Y, McIntire E J B, Boisvenue C, et al. Climatic change only stimulated growth for trees under weak competition in central boreal forests[J]. Journal of Ecology, 2020, 108(1): 36-46.
6. Aguirre O, Hui G, von Gadow K, et al. An analysis of spatial forest structure using neighbourhood-based variables[J]. Forest ecology and management, 2003, 183(1-3): 137-145.
7. Bhandari S K, Veneklaas E J, McCaw L, et al. Investigating the effect of neighbour competition on individual tree growth in thinned and unthinned eucalypt forests[J]. Forest Ecology and Management, 2021, 499: 119637.
8. Bhandari S K, Veneklaas E J, McCaw L, et al. Individual tree growth in jarrah (Eucalyptus marginata) forest is explained by size and distance of neighbouring trees in thinned and non-thinned plots[J]. Forest Ecology and Management, 2021a, 494: 119364.
9. Forrester D I, Elms S R, Baker T G. Tree growth-competition relationships in thinned Eucalyptus plantations vary with stand structure and site quality[J]. European Journal of Forest Research, 2013, 132(2): 241-252.
10. Canham C D,LePage P T. A neighborhood analysis of canopy tree competition: effects of shading versus crowding[J]. Canadian Journal of Forest Research, 2004, 34(4): 778-787.
11. Liang H, Huang J G, et al. Contributions of competition and climate on radial growth of Pinus massoniana in subtropics of China[J]. Agricultural and Forest Meteorology, 2019, 274:7-17.
12. Stephenson N L, Das A J, Condit R, et al. Rate of tree carbon accumulation increases continuously with tree size[J]. Nature, 2014, 507(7490): 90-93.
13. Coomes D A, Flores O, Holdaway R, et al. Wood production response to climate change will depend critically on forest composition and structure[J]. Global change biology, 2014, 20(12): 3632-3645.
14. Roux P, Shaw J D, Chown S L . Ontogenetic shifts in plant interactions vary with environmental severity and affect population structure[J]. New Phytologist, 2013, 200(1).
15. Lhotka J M. Examining growth relationships in Quercus stands: An application of individual-tree models developed from long-term thinning experiments[J]. Forest Ecology and Management, 2017, 385: 65-77.

The Evaluation Model of Economy and Application in the Secondary Development by Real Options Method

Yao Huang

Toutai Oilfield Co., Ltd., Daqing, 163000, China

Abstract: The secondary development is a deeper development after efficient development, more difficult to develop, higher technical risks and fluctuations in oil prices all increase uncertainty. Owing to comprehensive benefits and uncertainty, it's essential to find a more flexibility in investment decision-making method instead of cash flow which may cause the deviations of investment decisions. Organic combining the evaluation model of capacity, risk analysis methods, Black—Scholes model, this article has established an option evaluation method of deliverability project which is fit for the secondary development, reflects comprehensive benefits and uncertainty. It has provided a more accurate method for investment decision on the secondary development evaluation under uncertainty.

Keywords: The secondary development; uncertainty; physical options; the economic evaluation; evaluation model.

1. Introduction

Secondary development is an emerging focal point aimed at enhancing the oil recovery rate in mature oilfields. It represents a significant strategic initiative and systematic engineering effort to elevate the level of oilfield development and enhance technological innovation capabilities. Planning and deployment of secondary development must adhere to principles of economic efficiency, as the economic evaluation conclusions are pivotal to the success or failure of project investment decisions. The unique aspects of secondary development, such as changes in well patterns, combinations of well networks and reservoir formations, and the integration of various displacement methods, render traditional economic evaluation methods and parameter determination approaches partially unsuitable for assessing secondary development. Additionally, the high uncertainty and limited known information associated with secondary development projects make conventional investment decision methods less flexible, potentially leading to investment decision errors. Therefore, it is imperative to establish an economic evaluation methodology that reflects the comprehensive benefits and uncertainties of secondary development projects.

2. The characteristics of economic evaluation for secondary development

2.1 The characteristics of secondary development projects

The secondary development of a joint-stock company refers to a strategic and systematic engineering initiative for mature oil (gas) fields with significant resource potential. When these fields are at a stage of low-speed, inefficient extraction under existing development conditions or are nearing abandonment, a novel conceptual approach and a "triple" technological roadmap are employed. This involves the reconstruction of a new development system aimed at substantially increasing the ultimate recovery rate of the oil (gas) fields. The primary objectives include achieving a strategic and comprehensive system engineering for safe, environmentally friendly, energy-efficient, and highly effective development [1].

The objective of secondary development is to significantly increase the recovery rate, primarily achieved through a series of practices aimed at improving oil recovery efficiency. These practices include enhancing displacement efficiency, implementing deep reservoir adjustments, and transitioning development methods. The characteristics of secondary development are as follows: First, Specific Targeting. It focuses on specific targets, namely mature oil fields with over twenty years of development (possibly fifteen years for heavy oil). The identified fields should have achieved a recovery level of 70% or more of the proven recoverable reserves, with an overall water cut exceeding 85%. Second, Complex Distribution of Remaining Oil. The remaining oil is distributed in a highly complex manner, making the identification of small oil layers and reserves a starting point, focal point, and challenging aspect of secondary development. Third, Comprehensive Multi-Objective Approach. This involves considering multiple objectives: 1. Comprehensive consideration of geological characteristics, remaining oil distribution, reservoir dynamics, oil recovery processes, and surface facilities for a systematic, all-encompassing, and deep-level transformation of mature oil fields. 2. Multiple objectives include increasing reserves, boosting production, energy conservation, reducing consumption, ensuring safety, and maintaining environmental protection.

2.2 The characteristics of economic evaluation of secondary development compared with conventional development method

The economic evaluation of secondary development construction projects is a crucial component of the preliminary research phase for such projects. It should be established on the foundation of reservoir engineering, drilling engineering, oil recovery engineering, and surface engineering optimization. Utilizing scientific and standardized analytical methods, the economic evaluation aims to substantiate the economic feasibility of secondary development projects in mature oil fields. This process provides a basis for informed decision-making.

Analyzing the technical aspects of various specialized engineering projects in secondary development, although each engineering technology possesses distinct characteristics, the fundamental nature of the project lies in mobilizing increments from existing reserves with relatively minimal additional inputs to achieve substantial additional benefits [2]. It exhibits typical characteristics of a renovation and expansion project. Therefore, in the economic evaluation, the "incremental benefits" metric should be applied to assess the project's economic viability. As secondary development involves a profound development following efficient exploitation, it is rooted in the reconstruction of the subsurface knowledge system, restructuring of well networks, and reorganization of surface processes. Its focus is on the sustainable development of mature oil fields. Consequently, unlike typical oilfield renovation and expansion projects, its benefits should be comprehensive, necessitating evaluation from multiple levels and perspectives [3]. Multilevel and multi-perspective evaluation entails examining the project's incremental benefits within the scope of the mature oil field, assessing overall incremental benefits within the entire oil region from the perspective of the oilfield branch, and calculating comprehensive incremental benefits within the entire business scope from the viewpoint of the group company. The economic evaluation of secondary development construction projects must ensure objectivity, scientific rigor, and fairness. It involves a combination of quantitative and qualitative analyses, with a principle emphasizing quantitative analysis, and a combination of dynamic and static analyses, primarily favoring dynamic analysis. This approach is crucial for evaluating the benefits brought about by investments in secondary development over time.

Due to the widespread application of new technologies, the intricate distribution of various reservoirs, the complexity of stratigraphic well networks, and increased uncertainty in oil prices, the risk associated with secondary development has intensified. This renders traditional economic evaluation methods unsuitable for assessing secondary development. Instead, evaluation methods that can reflect uncertainties related to oil prices, technology, and other factors should be employed for a more accurate assessment.

Therefore, the characteristics of the economic evaluation methods for secondary development can be summarized as follows: first, it reflects multiple benefits and is suitable for the technical characteristics of "triple one best". Second, it is suitable for the characteristics of the project with little known information, great risk and high uncertainty. The third is a flexible decision-making method.

3. The capacity project option valuation model for secondary development

3.1 The introduction of physical option value.

Conventional capacity project investment decision methods involve discounting cash flows (including NPV, IRR) [4]. This approach thoroughly considers the project's ability to generate future cash flows, providing a reliable estimation of expected returns. However, in investment decisions for projects like secondary development, where petroleum serves as a resource with inherent value, possessing such a resource implies ownership of that value regardless of its current development status. If it cannot be effectively implemented in the current period, investment in development may be viable in the future, contingent on changes in external conditions such as secondary development technologies and oil prices. To address the uncertainties and timing selection issues in secondary development investment projects, a new approach is required to determine the project's contingent value.

A real option is the right to choose among physical assets, giving the option holder the ability to adapt their investment behavior based on the development trends of specific entities [5]. It allows the holder to make the most appropriate decision at the most opportune time. According to the fundamental idea of real option analysis, the actual value of an investment project's real option is equal to the Net Present Value (NPV) of the project plus its flexibility value. Therefore, the real value of a project is also known as the expanded Net Present Value (NPVN), distinguishing it from the traditional NPV. The value of the opportunity is also referred to as the option value (C). The real options approach is suitable for investment decisions with long cycles, high investment costs, and significant investment risks. Using real option pricing models, one can accurately estimate the opportunities faced during decision-making. Moreover, based on the factors influencing these opportunities, one can actively adjust, control, or enhance the value of opportunities to minimize risks and improve strategic investment decisions. The real options theory breaks free from the constraints of traditional decision analysis methods. It doesn't simply negate traditional methods like NPV [6]; instead, it extends them by responding positively to uncertainty factors and their corresponding environmental changes, providing an extension of traditional methodologies.

The determination of the real option value C can be achieved using the Black-Scholes option pricing model [7], a primary model introduced in the early 1970s by Fisher Black and Myron Scholes. The Black-Scholes model is the predominant option pricing model in the field of real options. The formula is expressed as follows:

$$C = SN(d_1) - Ke^{-rt} N(d_2) \tag{1}$$

In which:

$$d_1 = \frac{log(S/K) + rT + \sigma^2 T/2}{\sigma\sqrt{T}}$$

$$d_2 = \frac{\log(S/K) + rT - \sigma^2 T/2}{\sigma\sqrt{T}}$$

In the formulas:

C represents the option value of the investment project.

S is the current price of the underlying asset, i.e., the present value of the expected cash flow returns from the project.

K is the exercise price of the option, i.e., the expected investment cost of the project.

T denotes the expiration date of the option, i.e., the duration of the investment opportunity for the project.

r is the risk-free compounded interest rate.

σ^2 is the price volatility, i.e., the variance of the annual compounded return rate.

$N(d_1)$ is the cumulative probability distribution function of the standard normal distribution.

$N(d_2)$ is the cumulative probability distribution function of the standard normal distribution.

3.2 Model of Uncertain Value for Secondary Development Projects

In the context of oil field development and construction projects, it is fundamentally suitable to utilize option valuation models to determine the uncertain value of the project. When applying option pricing models to value natural resource options, a crucial consideration is the impact of the lag in resource extraction on the option's value. Since natural resources cannot be extracted immediately, there is a time gap between the decision to extract and the actual extraction. Therefore, if there is a one-year lag in development potential, the current price of the potential should be discounted for one year, using the discount rate based on the "cash flow/asset value" ratio (dividend yield). The adjustment method for the temporal nature of extraction is to discount the current value of the potential based on the length of the lag period.

The real options model is directly derived from financial options [8], but it deals with tangible assets. Tangible assets have many characteristics distinct from financial assets, and within tangible assets, there are often multiple options that interact. Therefore, the valuation of real options is more complex than financial options, presenting numerous challenges that need to be addressed.

In the context of oil field development and construction projects, conditions conducive to utilizing option valuation models for estimation include [9]:

(1) The underlying asset is a development block about to be invested in for capacity construction. The block's price is determined by the development value of the block, and its fluctuations follow a continuous process.

(2) The variance can be determined using a risk analysis method, hence it is known.

(3) For oil field development and construction projects with development potential, they can be immediately put into development. Therefore, the option can be exercised immediately. One of the premises for the validity of option pricing models is that the option can be exercised immediately.

Generally, the value of an option is primarily influenced by five factors: the current price of the underlying asset, volatility, the option's exercise price, time to expiration, and the risk-free compounded interest rate. When combined with the application of real options and financial options, the meanings of these parameters in the context of oil field development and construction project options should be:

(1) Current Price of the Underlying Asset: The present value of the expected cash flow returns from the currently undeveloped capacity construction block (S), $S = \sum_{t=1}^{n} CI_t (1+i_c)^{-t}$, where CI_t represents the expected cash flow returns for each year of the evaluation period.

(2) Volatility: The fluctuation in project value due to the volatility of oil prices and deviations in reserve estimates (σ^2).

(3) Exercise Price of the Option: The development cost of the oil field project (K), $K = \sum_{t=1}^{n} L\&C_t (1+i_c)^{-t}$, where $L\&C_t$ includes investment and cost expenditures during the evaluation period.

(4) Time to Expiration: The expiration date of the option during the rights period (T).

(5) Risk-Free Compounded Interest Rate: The risk-free interest rate (r).

According to the above analysis of option valuation model parameters for capacity construction, the option valuation model for capacity construction can be expressed by the following formula:

Call Option:

$$C = \sum_{t=1}^{n} \left[CI_t \cdot (1+i_c)^{-t} \right] \cdot N(d_1) - \sum_{t=1}^{n} \left[L\&C_t \cdot (1+i_c)^{-t} \right] \cdot e^{-r} \cdot N(d_2) \tag{2}$$

Put Option:

$$C=\sum_{t=1}^{n}\left[CI_t\cdot(1+i_c)^{-t}\right]\cdot\left[N(d_1)-1\right]-\sum_{t=1}^{n}\left[L\&C_t\cdot(1+i_c)^{-t}\right]\cdot e^{-rT}\cdot\left[N(d_2)-1\right] \quad (3)$$

In which:

$$d_1=\frac{\ln(\sum_{t=1}^{n}CI_t(1+i_c)^{-t})-\ln(\sum_{t=1}^{n}I\&C_t(1+i_c)^{-t})+(r+\frac{\sigma^2}{2})T}{\sigma\sqrt{T}}$$

$$d_2=d_1-\sigma\sqrt{T}$$

Under uncertain conditions, the value of a project is determined by both its intrinsic value and option value. According to the option valuation model for capacity construction, the potential development value (C) at different times T can be calculated. This value, added to the net present value (NPV) of the overall project benefits, provides the value of the project considering uncertainty for the secondary development specialized technology capacity construction project. Therefore, the model representing the total value under uncertainty is expressed as:

$$NPVN = NPV + C \quad (4)$$

In the formulas: NPVN is the total project value after considering uncertainty, in thousands of yuan.
When the Net Present Value (NPV) of the overall project benefits, NPV < 0 and NPV + C≥0, the project can be deferred for development until period T. If NPVN < 0, the project is currently not acceptable.

4. Application of the Uncertain Value Model for Secondary Development Projects

4.1 Determination and Explanation of Evaluation Parameters

Evaluation of the secondary development block BBGDD, which involves the development of a low-viscosity oil reservoir through water flooding to steam flooding. During the assessment, the exchange rate is set at 6.8 RMB/USD, the crude oil volume conversion factor is 7.428 barrels/ton, and the oil price is $65 per barrel. The crude oil commercial rate is calculated at 98.37%. Tax-related evaluation parameters include income tax rate (25%), value-added tax rate (17%), resource tax (5%, with preferential rates according to documentation), city construction maintenance tax (7%), and education surcharge (3%). Cost data is forecasted based on current cost expenses and neighboring block conditions for the implementation of the secondary development.
In the option valuation parameters, the block's present value (S) and development investment (K) are measured by the above conventional evaluation calculation. The expiration period (T) represents the time until the block is planned for development, and for this evaluation, it is set at 1 year. The risk-free interest rate is the one-year Treasury bond rate (5%). The volatility of oil prices and the volatility of reserve estimates (variance σ^2) are determined through risk analysis of the block.

4.2 Evaluation Results

Steam drive technology is currently predominantly applied in heavy oil reservoirs, while steam drive development in light oil is in its initial exploration phase, with substantial technical risks. Accurately assessing its development benefits can provide a basis for informed decision-making.
Through sensitivity analysis, it is found that the economic benefits of the transition zone steam drive are significantly influenced by uncertain factors such as Enhanced Oil Recovery (EOR) and steam prices (Figure 3-1). Under current conditions, the cost of steam for the project is 180 yuan per ton. According to the scheme analysis, BBGDD can increase the recovery rate by 17.89%. The project, evaluated using conventional methods even after considering comprehensive benefits, achieves an internal rate of return (IRR) of only 3.39%. Considering the potential widespread adoption of steam drive technology, if the steam acquisition price drops to 150 yuan per ton, the internal rate of return, considering multiple benefits, is 8.95%, still below the 12% standard, with a net present value of -21.86 million yuan. However, considering uncertain factors such as future technological advancements and oil price changes, the project has an option value (potential profitability) of 36.5 million yuan. The net present value of the project then reaches 14.64 million yuan (Table 3-1).

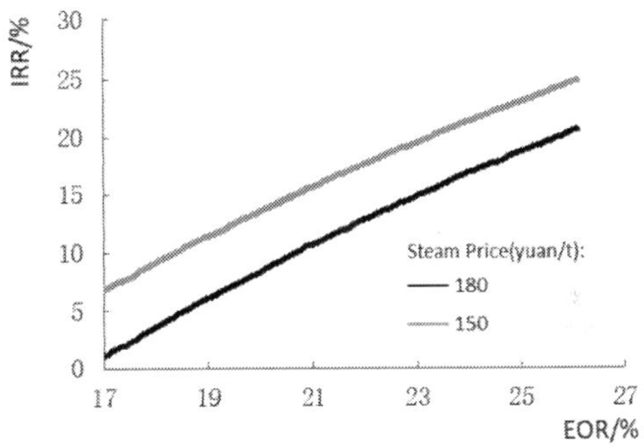

Figure 3-1 Benefit Variation Chart at Different Recovery Rates

Although the internal rate of return (IRR) indicator for the BBGDD block remains below 12% even when considering comprehensive benefits, the project, with its potential development value under uncertain conditions, can be planned and arranged as a valuable potential block.

Table 3-1 Economic Evaluation Results Table for BBGDD under Different Conditions

Oil Price: $65 per Barrel

project		180			150		
	Steam price (yuan/t)	180			150		
	EOR (%)	17.89	21.62	24.11	17.89	21.62	24.11
IRR (%)	Conventional Project Benefits	-0.13	7.66	12	5.08	12.02	16.11
	Comprehensive Benefits	3.39	12	16.98	8.95	16.76	21.51
Total Value Considering Options (in thousands of yuan)		-4485.6	5032.8	11926.2	1464.3	11734.6	19058.5

5. Conclusion

(1) In the investment decision-making of oilfield secondary development, the inherent value of petroleum resources, regardless of their current development status, is a crucial factor. Possessing such resources means possessing their inherent value. The incorporation of real options can effectively reflect the latent value of resources. The establishment of an economic evaluation model for secondary development considering real options deviates from the traditional "with-or-without" net present value approach, enabling a comprehensive assessment. This model addresses the limitations of not capturing integrated benefits and dealing with uncertainties arising from new technology choices, oil price fluctuations, and other risks. It is well-suited for investment decisions in current uncertain conditions for secondary development projects.

(2) The real options-based economic evaluation model for secondary development is suitable for investment decisions at different time periods within the current management system. It enhances the scientific, standardized, and flexible aspects of investment management, holding significant importance for evaluating the secondary development of mature oilfields.

References

1. Hu W. Necessity and feasibility of PetroChina mature field redevelopment[J]. Petroleum Exploration and Development, 2008, 35(1): 1-5.
2. Xu Jianjun, Huang Lida, Yan Limei, Yi Na. Insulator Self-Explosion Defect Detection Based on Hierarchical Multi-Task Deep Learning[J]. Transactions of China Electrotechnical Society, 2021,36(07):1407-1415.
3. Yan Limei,LIU Yongqiang,XU Jianjun,et al.Broken string diagnosis of composite insulator based on Grabcut segmentation and filler area discrimination[J].Power System Protection and Control,2021,49(22):114-119
4. Yi, Q. Wang, L. Yan, et al., A multi-stage game model for the false data injection attack from attacker's perspective. Sustainable Energy Grids & Networks 28 (2021).
5. Na Yi,Jianjun Xu,Limei Yan,Lin Huang. Task Optimization and Scheduling of Distributed Cyber-physical System Based on Improved Ant Colony Algorithm. Future Generation Computer Systems, 109(Aug. 2020),134-148.
6. Yang Zhao, Jianjun Xu, Jingchun Wu. A New Method for Bad Data Identification of Oilfield Power System Based on Enhanced Gravitational Search-Fuzzy C-Means Algorithm. IEEE Transactions on Industrial Informatics. VOL. 15, NO. 11, NOVEMBER 2019 5963-5970
7. Jing Han,Xi Wang,LiMei Yan, Aida Dahlak, et al. Modelling the performance of an SOEC by optimization of neural network with MPSO algorithm. International Journal of Hydrogen Energy, Volume 44, Issue 51, 22 October 2019, Pages 27947-27957. https://doi.org/10.1016/j.ijhydene.2019.09.055
8. Zhang Yongfeng, et al. Real option characters of petroleum exploration and development projects[J]. Petroleum Exploration and Development, 2006, 33(1): 115-118.
9. Li R, et al. Evaluating model and its application of the productivity project under uncertain condition[J]. Petroleum Geology & Oilfield Development in Daqing, 2012,31(1):66-69.
10. Zhao Xiaoguang. Improvement of project economic evaluation method by real option theory[J]. Market Modernization, 2008, (03): 160-161.

Analysis of heat and mass transfer characteristics of supercritical CO_2 in vertical U-tube

Chenhong Bai[1,*], Ying Huang[1], Xinjun Li[2]

[1]Shanghai Aircraft Design and Research Institute, Shanghai, China
[2]School of Energy and Mechanical Engineering, Nanjing Normal University, Nanjing, China

Abstract. The heat and mass exchange of CO_2 in a vertical U-tube under supercritical pressure was simulated. The effects of the geometry of the bend area, The center arc radius of the bend area and the different bend orientations on the flow heat transfer of supercritical CO_2 in the U-tube were analyzed. The results show that in the bend area, the flow is affected by the combined action of centrifugal force and gravity, as well as the pipe structure, so the flow mixing is intense, and the heat transfer capacity is significantly enhanced. When the direction of the bend changes, the interaction between centrifugal force and gravity will affect the heat transfer at pipe bending. When the central arc radius r of the bend area increases from 0.5D to 3D, the heat transfer at the bend can be strengthened, but will weaken the heat transfer at the bend outlet area.

1 Introduction

As a new and efficient refrigerant, supercritical fluid has been widely used in chemical engineering, power engineering, aerospace technology and other fields. Due to the drastic change of the thermal properties of supercritical fluid in the pseudo-critical zone, the study of its flow heat transfer in various pipelines has received more and more attention. The use of supercritical fluid as refrigerant is a promising new refrigeration method. However, due to the complex change of thermal properties of supercritical fluid, numerous constraints of supercritical heat transfer experiment, and various heat transfer situations in pipelines, supercritical heat transfer still needs to be studied in depth. At present, the research of supercritical fluid working medium at home and abroad mainly focuses on water, CO_2, helium and nitrogen[1]. The research on supercritical fluid flow pipeline at home and abroad mainly focuses on straight passage, U-shaped passage, serpentine passage and spiral passage. U-shaped channel is a kind of structure widely used in heat exchange equipment. Because the flow of ordinary fluid will turn when flowing through the U-shaped bend area, the flow will be separated to form a secondary flow, so that the local pressure drop[2] can not be ignored. For supercritical CO_2, the conditions will be more complicated, because it is subject to the combined action of buoyancy, gravity and centrifugal force. The complex flow and heat transfer characteristics of supercritical CO_2 in the U-tube will determine the overall performance of supercritical CO_2 heat exchangers. Therefore, this paper intends to study the flow heat transfer of supercritical CO_2 in the U-tube, and analyze the influence of the number of bends and the geometry of the bends on the flow and heat transfer characteristics of supercritical CO_2, which provides a partial reference for the design of efficient supercritical CO_2 heat exchanger.

Zhang Lina et[3] al., Wang Junhui et[4] al., studied the heat transfer characteristics of horizontal circular tubes, and analyzed the influence of heat flux, wall temperature, mass flow rate and pressure on the convective heat transfer of CO_2. Jiang et[5] al. conducted experiments and numerical simulations on convective heat transfer of supercritical CO_2 in a vertical circular tube, and the results showed that flow direction and velocity had little influence on heat transfer, and no heat transfer deterioration occurred. Wei et[6] al., Ma Xu et[7] al studied supercritical flow heat transfer in a rotating U-tube, and analyzed the influences of rotational speed, rotational radius and pipe diameter on heat transfer characteristics. Xu et[8] al. conducted an experimental study on the characteristics of carbon dioxide heat transfer in the spiral tube, and found that the fluid turbulence was increased when flowing in the spiral tube, and the influence of floating lift and gravity on the fluid heat transfer factor was intensified, so the heat transfer efficiency of the fluid in the spiral tube was higher than that in the straight tube. Nemati et[9] al. found that, compared with ideal gas heat transfer, the additional term in the average flow control equation has a significant effect on the energy balance. The Reynolds mean Navier-Stokes equation (RANS) is an economical and easy way to model turbulence. Pecnik et[10] al. point out that multiple revisions to turbulence models often negatively affect the accuracy of the model's turbulent kinetic energy. He et[11,12] al. carried out numerical studies using the two-path turbulence model and found that the turbulence model significantly overestimated the flow stratification, which led to the deterioration of heat transfer.

* Corresponding author:baichenhong@comac.cc

Based on the continuity equation, momentum conservation equation and energy conservation equation of compressible fluid, and considering the thermophysical parameters of supercritical CO_2 as well as the compressibility effect and floating lift effect of the flow, this paper studies and analyzes the heat and mass transfer characteristics of supercritical CO_2 when it flows in a U-tube uniformly heated on the outer wall by numerical simulation. And through the distribution characteristics of the flow field velocity, temperature, turbulent kinetic energy and heat transfer coefficient inside the tube to further analyze the heat and mass transfer characteristics of supercritical CO_2 flowing in the U-tube. And the special phenomena of fluid passing through U-tube and the influence of the change of center arc radius in bend area on the heat and mass transfer ability of fluid flowing in tube are analyzed.

2 Governing equation and physical model

2.1 Governing Equations

The governing equation[13] for fluid flow heat transfer in a U-tube heated with uniform heat flux is as follows:
Continuous equation:

$$\frac{\partial(\rho u_i)}{\partial x_i}=0 \tag{1}$$

Momentum equation:

$$\frac{\partial(\rho u_i u_j)}{\partial x_j}=-\frac{\partial p}{\partial x_i}+\frac{\partial}{\partial x_j}\left[\mu\left(\frac{\partial u_i}{\partial x_j}+\frac{\partial u_j}{\partial x_i}\right)\right]-\frac{2}{3}\frac{\partial}{\partial x_i}\left(\mu\frac{\partial u_j}{\partial x_j}\right)+\rho g_i$$
(2)

Energy equation:

$$\rho c_p u_j\frac{\partial T}{\partial x_j}+\rho u_j\frac{\partial\left(-\frac{1}{2}\omega^2 r^2\right)}{\partial x_j}=\frac{\partial}{\partial x_j}\left(k\frac{\partial T}{\partial x_j}\right)+T\beta w_j\frac{\partial p}{\partial x_j}+\Phi^* \tag{3}$$

Where $\Phi^*=\frac{\mu}{2}\left(\frac{\partial u_i}{\partial x_j}+\frac{\partial u_j}{\partial x_i}\right)^2-\frac{2}{3}\mu\left(\frac{\partial u_i}{\partial x_i}\right)^2$ 。

In the above formula, x is the Cartesian coordinate,m; i is the general spatial index; g is acceleration of gravity, m/s^2; r is the radius of curvature of the bend, m; u, w are radial and axial velocity respectively, m/s; p is fluid pressure, Pa; T is the fluid temperature, K; cp is the specific heat capacity of the fluid, J/(kg·K); ρ is fluid density, m/s^2; μ is hydrodynamic viscosity, Pa·s; λ is the fluid thermal conductivity, W/(m·K); β is the coefficient of volume expansion, K^{-1}; k is the turbulent kinetic energy, m^2/s^2.

2.2 Physical model and boundary conditions

In this paper, the flow and heat transfer characteristics of supercritical CO_2 in U-tube are numerically simulated. The physical model is shown in Fig. 1.

The flow area of supercritical CO_2 adopts a double-bend U-tube model. The pipe is vertically arranged as shown in the figure. Supercritical CO_2 with an inlet temperature of 25°C Tin enters from the bottom of the flow inlet and flows vertically upward into a straight pipe section with inner diameter D of 9mm. When flowing in this section of pipe and subsequent pipes, the fluid is uniformly heated by a heat flux q of 16 kw/m^2 outside the pipe. After passing through the 500mm straight pipe section, it enters the U-shaped bend, which is a bend with an arc upward, and the diameter of the pipe with three times the central arc radius of the bend

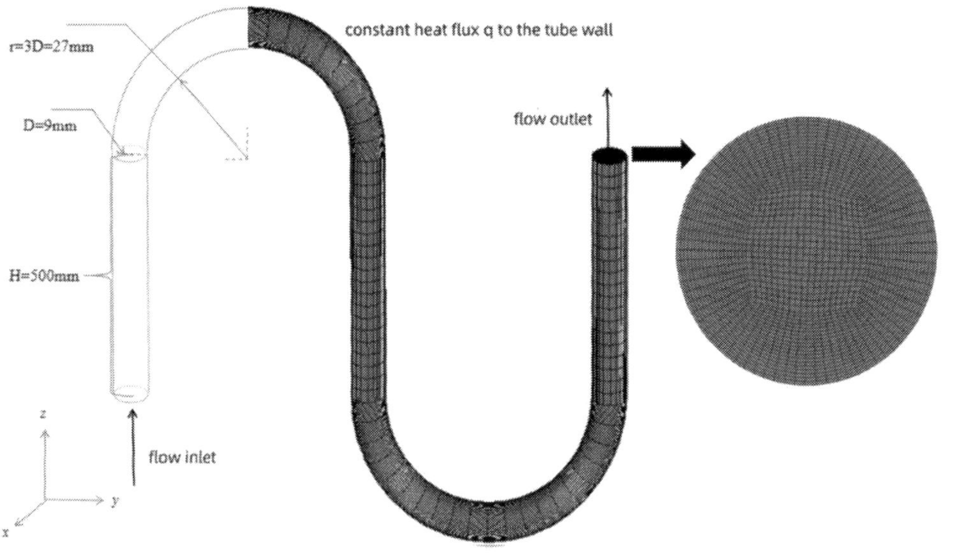

Fig. 1. U-tube physical model and grid division

section is 27mm, and then passes through the straight pipe with the same size and then enters a U-shaped bend with the arc downward, the size of the pipe is unchanged, and finally flows out through the straight pipe upward. The mass flow rate of supercritical CO_2 flowing in the pipe is 300kg/m²·s.

The above physical model is divided into meshes. In order to make the simulation results closer to the experimental results, the U-tube is divided into meshes as shown in Fig. 1. Hexahedral structured meshes are adopted, uniform meshes are adopted along the fluid flow direction, non-uniform meshes are adopted in the radial direction, and the size of the meshes is smaller the closer it is to the wall. The O-shaped grid is used to divide the cross-section of the round pipe, which can solve the problem that the distortion rate of the grid in the center of the cylinder is too large, and the grid division at the bend of the pipe is shown in Fig. 1.

3 Result analysis and discussion

In this paper, ANSYS FLUENT is used for calculation, SST k-ω[14] model is used for turbulence model, and curvature correction is carried out, and SMPLIC algorithm is used for pressure velocity coupling.. Since the thermal properties of CO_2 change rapidly near the pseudo-critical temperature, the model adopts the variable properties. Based on the self-programming of REFPROP in NIST database, the supercritical CO_2 thermal properties journal file is established in advance. During calculation, the journal file is directly read in FLUENT to obtain the thermal physical property parameters such as ρ, cp, λ and μ. The operating pressure of the whole simulation experiment is P=8MPa, the operating temperature is T=25°C, and the acceleration of gravity in the z direction is gz=9.8N/kg. In this paper, the turbulence intensity is defined as the ratio of the root mean square of velocity fluctuation to the average velocity. The boundary conditions of the example are set as shown in Table 1.

Table 1. Example boundary conditions

surface	type	Parameters	Numerical value
Imports	Mass flow rate imports	Inlet mass flow rate（kg/m²·s）	300
		Inlet gauge pressure（MPa）	~
		Inlet temperature（°C）	25
exit	Pressure outlet	Outlet gauge pressure（MPa）	~
		Outlet temperature（°C）	~
Pipe wall	Walls	Pipe wall heat flow per unit area（kW/m²）	16
Pipe geometry		Pipe diameter D(mm)	9
		Single side straight pipe length H（mm）	500
		Bend section center arc radius r（mm）	27
Operating pressure P=8MPa, operating temperature T=25 degrees Celsius, gravity is g_z=-9.8N/kg			

3.1 Model verification

In order to ensure the accuracy of the calculation model, according to the physical model and method described above, the heat and mass transfer in supercritical tubes was verified by using the experimental results of Dang and Hihara[15] in literature [15]. The parameters of the verified model are inlet temperature Tin=25°C, inlet pressure Pin=8MPa, tube length L=475mm, tube inner diameter D=6mm, inlet mass flow rate G= 200,400 kg/(m²·s), wall heat flux q=12.0kW/m2.

The comparison between numerical calculation results and experimental data is shown in Fig. 2. It can be seen that the calculated results are basically consistent with the experimental results of the two mass flow rates. When the mass flow rate is G=200kg/(m²·s), the numerical calculation results are consistent with the experimental data. When G=400kg/(m²·s) and the average CO_2 temperature is 308-316°C, the numerical simulation average heat transfer is significantly higher.

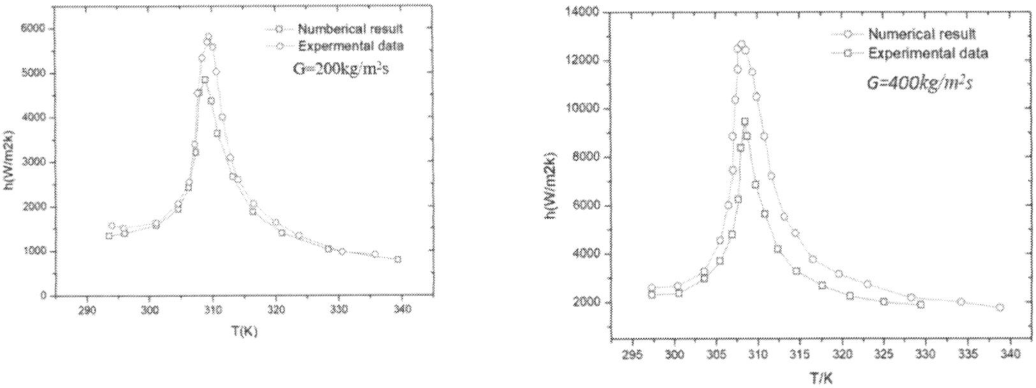

Fig. 2. The relationship between the average heat transfer coefficient and the mainstream temperature. (Source of experimental data [17])

This may be due to the fact that the temperature of CO_2 is near quasi-critical when the tube is flowing. In order to verify the validity of the numerical method, the flow and heat transfer characteristics of supercritical CO_2 in horizontal tubes were simulated using the same grid scale and numerical method, and compared with the published experimental results.

Before using the model for simulation calculation, it is necessary to verify the grid independence and select grid types with similar fit degree and relatively small number of grids. In this paper, three structured grids with different distribution types are created, which mainly change the grid distribution in the radial way and distribute the grid evenly in the axial way. Specific parameters are shown in Table 2. Fig. 3 shows the simulation results of the three types of grids under the conditions set above. There is a small difference in the calculation results of the three grids, among which Case 2 and Case 3 basically coincide, which indicates that Case 2 not only meets the need of independence, but also meets the calculation accuracy requirements. Therefore, the grid in Case 2 is selected in this paper for the subsequent numerical simulation analysis.

Table 2 describes the grid division status

Grid Types	Radial network management Distribution	Axial grid distribution	Total grid number	Mesh quality
Case 1	15×15×5	100-50-100-50-100	404609	0.75~1
Case 2	20×15×5	100-50-100-50-100	739444	0.77~1
Case 3	20×20×5	100-50-100-50-100	1174529	0.8~1

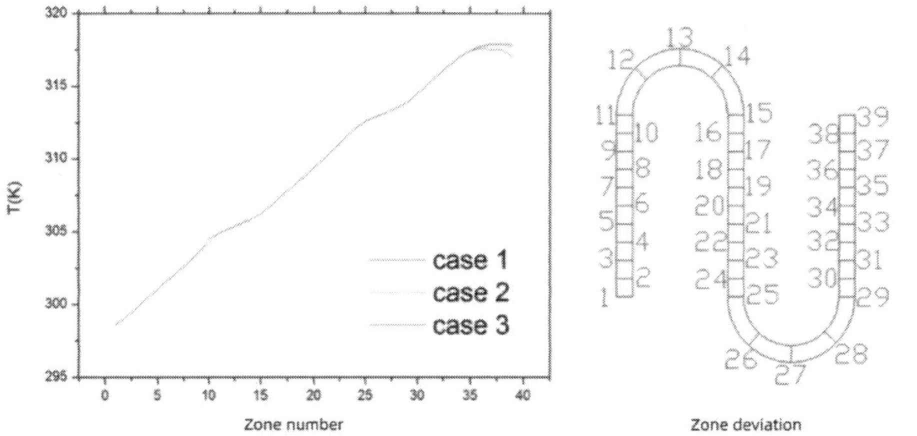

Fig. 3 Comparison of calculated values of mean temperature of cross section

3.2 Influence of the geometry of the bend area on the flow and heat transfer characteristics

In this paper, the flow heat transfer characteristics of supercritical CO_2 in U-tube are studied. Fig. 4 and Fig. 5 show the distribution of temperature and flow velocity at different sections in the U-tube.

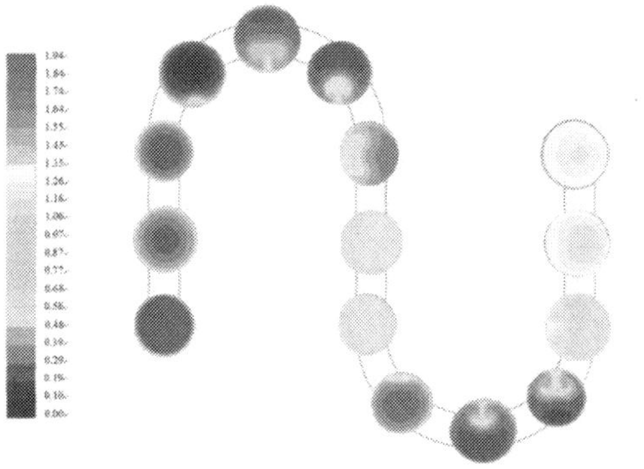

Fig. 4 Temperature distribution of U-tube section

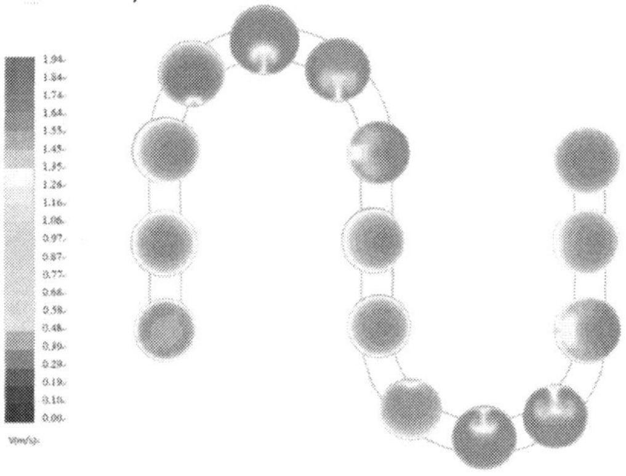

Fig. 5 Velocity distribution of U-tube section

It can be found that when supercritical CO_2 flows vertically upward in the straight pipe at the front end of the U-bend, its internal fluid temperature distribution is exactly the same, and due to the heating effect of the tube wall, the fluid temperature increases gradually from the entrance, forming a uniform high-temperature fluid zone around the wall, and the mainstream temperature is relatively low. When the conventional fluid enters the U-bend area, the mainstream fluid will shift to the outer wall due to the action of centrifugal force, and the flow of the fluid will no longer be one-way flow, forming a secondary vortex structure, called Dean vortex. However, since the density of supercritical CO_2 changes dramatically with temperature near the quasi-critical point, the gravity change dominates when supercritical CO_2 flows through the bend area. The mainstream of high-density fluid sinks towards the inner wall under the action of gravity, forming a secondary vortex structure opposite to the Dean vortex form, as shown in the cloud image of the bend area in Fig. 5. After flowing out of the bend area, the non-uniform temperature field and velocity field gradually dissipate and weaken with the mixing of the fluid. However, at this time, due to the influence of pressure gradient, the fluid will accelerate separation at the outlet of the bend area, forming a low-speed and high-temperature fluid area.

Fig. 6 shows the distribution of turbulent kinetic energy at different sections of U-tube. Before entering the U-shaped bend area, the overall level of turbulent kinetic energy is low, and the floating lift near the wall is relatively large. When the fluid enters the U-bend area, the fluid with higher turbulent kinetic energy begins to sink and move to the inner wall under the action of gravity. Due to the separation of the fluid, the turbulent kinetic energy of the fluid after the bend is much greater than that before the bend. The increase of turbulent kinetic energy at the bend enhances the heat transfer performance of the fluid. Therefore, the study of the bend area plays an important role in improving the heat transfer performance of heat exchange equipment.

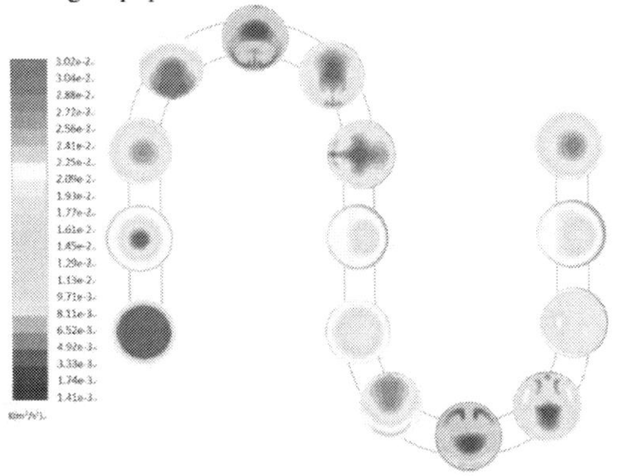

Fig. 6 Turbulent kinetic energy distribution of U-tube section

3.3 Influence of central arc radius of curved pipe on flow and heat transfer characteristics

In order to analyze the influence of the central arc radius of the bend on the flow and heat transfer, based on the above examples, the flow heat transfer characteristics of the bend section with the central arc radius of 0.5D=4.5mm, 1.5D=13.5mm and 3D=27mm are calculated in this paper. The change of the average wall temperature with the flow path of the fluid in the tube and the change of the average heat transfer coefficient with the central arc radius of the bend are shown in Fig. 7.

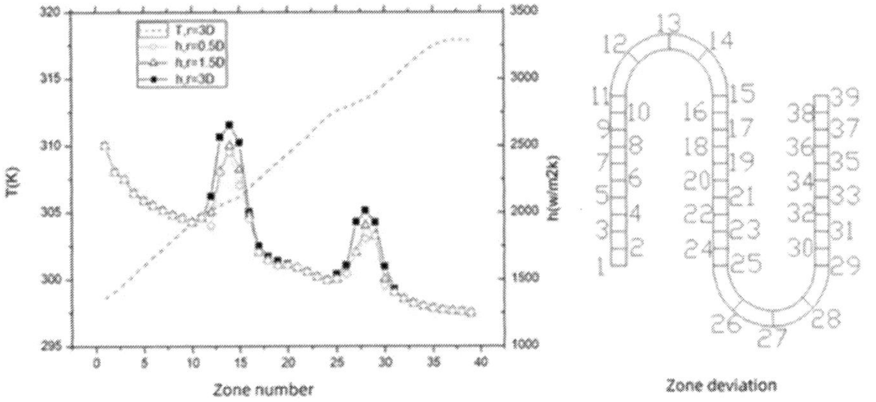

Fig. 7 Average heat transfer coefficient distribution of U-tube with center arc radius of different bending sections

There are two abrupt changes in the heat transfer coefficient in the curve, because the larger the central arc radius of the bend, the larger the centrifugal force, and the increase of the centrifugal force will lead to the increase of radial velocity, thus strengthening the secondary flow at the section and enhancing the heat transfer capacity. However, as the central arc of the bend increases from 0.5D to 3D, the flow separation and flow mixing strength weaken, and the fluid with high turbulent kinetic energy moves from the inner wall to the outer wall, which gradually reduces the level of turbulent kinetic energy and weakens the heat transfer ability. Therefore, as shown in Fig. 7, when the radius of the central arc of the bend is changed, the overall heat transfer capacity does not change greatly, and there are certain differences in the average heat transfer coefficient in the bend area. It can be judged that there is only one optimal bend center arc radius for U-tubes with different pipe diameters.

4 Conclusion

In this paper, the heat and mass transfer characteristics of supercritical CO_2 in U-tube are studied by numerical simulation. The influence of different bending center arc

radius and bending direction on supercritical temperature field is compared. The following conclusions are obtained:
(1) After CO_2 absorbs heat in the pipe, its temperature increases gradually along the inlet to the outlet, most of the fluid pressure in the pipe exceeds the critical pressure, when the temperature reaches the critical temperature, the fluid will enter the supercritical state. When supercritical CO_2 is in the bend area, it is affected by gravity to form a secondary flow structure, which is opposite to the Dean vortex of conventional fluid. After the fluid flows out of the bend area, the flow separation and secondary flow lead to the sharp fluctuation of the pressure in the bend area.
(2) As the radius of the center arc of the bend increases from 0.5D to 3D, the centrifugal force increases during the bend. The increase of centrifugal force can strengthen the secondary flow of fluid flowing in the tube, and play a role in strengthening the heat transfer at the bend. After turning, the flow separation and fluid mixing weaken with the increase of the bend center arc radius, and the fluid will separate and form a low-speed and high-temperature fluid to move to the outer wall after flowing out of the bend, weakening the heat transfer of the fluid in the straight tube after the turn.
(3) When U-tube is used in supercritical CO_2 heat exchanger, the heat transfer characteristics of U-tube can be optimized by changing the geometric parameters and layout of U-tube. There are many factors that can be optimized and improved in future experiments and studies to further enhance the heat exchange performance of supercritical heat exchangers.

References

1. M. Monjurul Ehsan, Zhiqiang Guan ,et al. A comprehensive review on heat transfer and pressure drop characteristics and correlations with supercritical CO_2 under heating and cooling applications, Int J. Renewable and Sustainable Energy Reviews, (2018),**92**:658–675.
2. Xu. Quanyong, Yantao Yang, Reynolds stress constrained large eddy simulation of separation flows in a U-duct, Propul. Power Res. (2014),3(2):49-58.
3. Zhang Lina et al. .Analysis of flow and heat transfer characteristics in supercritical CO_2 tube[J].Journal of Zhengzhou University, (2009), 30(4):69-72.
4. Wang Junhui, Guo Pengcheng, et al. Experimental studies on convective heat transfer of supercritical carbon Dioxide in a horizontal circular mini-tube [J]. Journal of Xi 'an University of Technology, (2018), 34(3):272-277.
5. JIANG P X,LIU B,ZHAO C R, et al. Convection heat transfer of supercritical pressure carbon dioxide in a vertical micro tube from transition to turbulent flow regime. Int. J. Heat Mass Transf. (2013),56(1-2):741-749.
6. Wei Chen, Zenan Yang, et al. Numerical investigation of heat transfer and flow characteristics of supercritical CO_2 in U-duct Int J. Applied Thermal Engineeringr, (2018), 144:532-539.
7. Ma Xu, Xie Guanghua, et al. Flow and Heat transfer characteristics of supercritical fluid in a Rotating U-Tube and Investigation of Influence Parameters.The 7th Chinese Society of Aeronautics and Astronautics Youth Science and Technology Forum, (2016),**7**:54-58.
8. Xu X X, LIU C B, et al. Experimental investigation on heat transfer characteristics of supercritical CO_2 cooled in horizontal helically coiled tube, Int J. International Journal of Refrigeration, (2016), **67**:190-201.
9. H. Nemati, A. Patel, B.J. Boersma, R. Pecnik, Mean statistics of a heated turbulent pipe flow at supercritical pressure, Int. J. Heat Mass Transf. (2015),83 :741–752.
10. R. Pecnik, A. Patel, Scaling and modelling of turbulence in variable property channel flows, Int J. Fluid Mech.(2017), 823.
11. S. He, W.S. Kim, J.H. Bae, Assessment of performance of turbulence models in predicting supercritical pressure heat transfer in a vertical tube, Int. J. Heat MassTransf.,(2008), 51 (19–20): 4659–4675.
12. S. He, W.S. Kim, J.D. Jackson, A computational study of convective heat transfer to carbon dioxide at a pressure just above the critical value, Appl. Therm. Eng.(2008), 28 (13):1662–1675.
13. Ma Xu, Zhang Tianyi, et al. Non-dimensional control equation of flow heat transfer in rotating U-tube and criterion parameter analysis.China Aerospace third professional information network 36th technical exchange conference, Gansu, Lanzhou, (2015).
14. S. Wang, W. Zhang, Z. Niu, J. Xu, Mixed convective heat transfer to supercritical carbon dioxide in helically coiled tube, CIESC J. 64 (2013) 3917–3926.
15. C Dang ,E Hihara, In-tube cooling heat transfer of supercritical carbon dioxide. Part 1:Experimental measurement, Int. J. Refrig.(2004),27(7),736-74.S

Research on the timing of polymer flooding oil and water well fracturing

Wei Zhang

No.5 Oil Production Plant, Daqing Oilfield Company Limited, Daqing 163513, Heilongjiang, China

Abstract: Fracturing technology is the main measure to increase production and injection in oilfield production. With the gradual expansion of the scale of polymer flooding, the proportion of fractured Wells also increases year by year. How to maximize the benefits of fracturing measures has become the key technology, and the timing of fracturing is a major factor affecting the fracturing effect. By studying the fracturing time and effect of fracturing Wells in the industrial zone of polymer flooding, the optimal fracturing time of injection-production Wells is given. The fracturing time of injection Wells should be dominated by blank water drive stage and water-cut decline period. The producing Wells with low liquid production strength and low flow pressure are suitable for whole-well fracturing in water-cut decline period, and the thick-injection-thin production Wells dominated by river channels are suitable for fracturing in low-water cut period. Thin and differential Wells with differential production are suitable for fracturing during water-cut recovery period.

Key words: polymer flooding; Fracturing; Injection-production well; Fracturing timing; Development phase

1. Introduction

In recent years, a total of 363 injection-production Wells (173 injection-190 production) have been implemented in the polymer flooding industrial zone, accounting for 23.4% and 21.5% of the total number of Wells, respectively. Among them, the fracturing proportion of Juxi block and Judong block injection Wells at the end of injection polymerization is 19.1% and 38.9%, and the fracturing proportion of produced Wells is about 40%. The proportion of fractured Wells in other blocks is gradually increasing.

2. Fracturing production Wells

According to the fracturing stage of the produced well, the fracturing effect is calculated according to the fracturing stage of the polymer flooding. The proportion of fracturing in low water cut period was 46.8%, and the fracturing proportion is higher, and the fracturing oil increase effect is the best, with the average oil increase per well reaching 3.5t. The fracturing ratio in water-cut down period and water-cut recovery period was 19.0% and 34.2%, and the average oil gain per well was about 2.5t.

Due to the strong timeliness of polymer flooding development, in order to maximize the fracturing effect, the fracturing timing of different injection stages is summarized by analyzing the effects of fracturing Wells that have been implemented.

2.1 Wells with low production fluid strength and low flow pressure are suitable for whole-well fracturing during water-cut decline period

This kind of well group is well developed and connected, but the strength of the produced liquid is less than 50% and the flow pressure level is less than 4.0MPa, and there is oil reservoir pollution. After polymer injection, the effect is slow, and fracturing can improve the fluid supply capacity of the whole reservoir, promote the polymer to advance in this direction, and guide the polymer flooding effect. The statistics of 33 Wells of this type include 28 Wells in water cut down period and 5 Wells in low water cut period. Compared before and after fracturing, there is no significant difference in daily oil increase between water cut down period and low water cut period, but the accumulative oil increase in water cut down period is better.

The results of numerical and modeling studies show (FIG.1) that for Wells with low liquid production strength and low flow pressure before 0.5PV, the earlier fracturing is performed, the greater the contribution of fracturing to the final recovery degree and the higher the final recovery degree; fracturing is performed after 0.5PV, and the contribution of fracturing to the final recovery degree is small. Therefore, this type of well is suitable for fracturing before 0.1PV in the water-cut decline period.

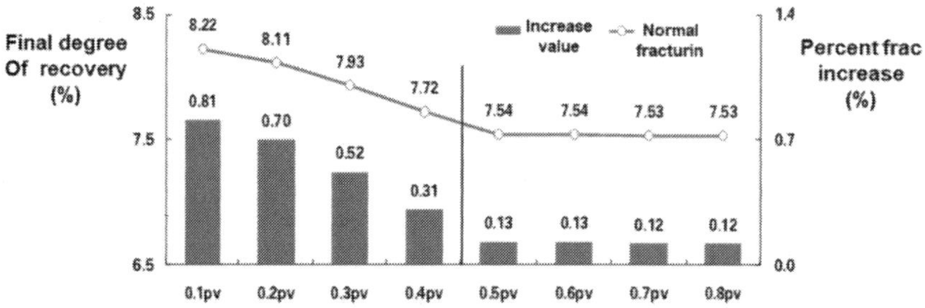

Fig. 1 Numerical model prediction of fracturing effect curves in different periods

2.2 The thick-injection-thin production Wells dominated by river channels are suitable for fracturing in the period of low water-cut value

According to the effect analysis data of 89 fracturing Wells in different types of well groups during the period of low water cut, it can be seen that the daily oil gain per well after fracturing of all kinds of well groups is more than 3.0t, among which the well group of thick injection and thin production type has the best effect, and the daily oil gain per well can reach 4.0t.

The results of the numerical modeling study show (FIG.2) that fracturing at different periods of Wells with thick injection and thin production can greatly improve the recovery degree, and the recovery degree can reach more than 10.0 percentage points. However, fracturing at around 0.2PV can achieve the highest recovery degree, reaching more than 11.0 percentage points.

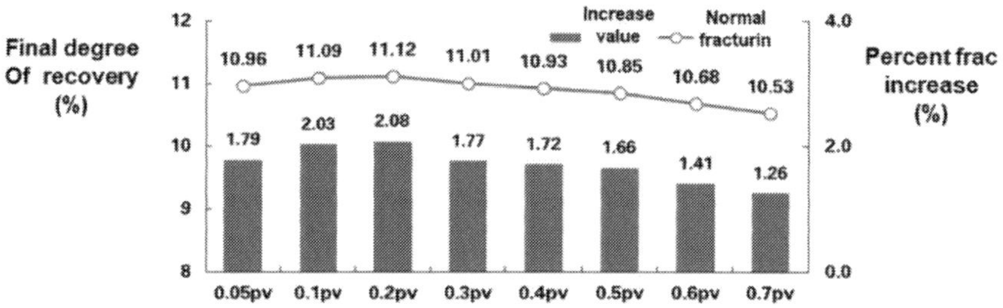

Fig. 2. Numerical model prediction of fracturing effect curves in different periods

As can be seen from the lacing curve of 28 thick-injection-thin production Wells after fracturing (FIG. 3), after fracturing in the low-water cut period, the liquid production increased, the water cut continued to decline, and the production and accumulation rate rose slowly. Through fracturing, the low-water cut period was extended, and better fracturing effect was achieved. The analysis shows that due to the good injection condition, the seepage capacity is effectively reduced, the sealing is better, and a good oil wall is formed in the low water content period. The producing end is reformed by fracturing to release the polymer flooding effect, and the water content continues to decrease after fracturing. Meanwhile, the further promotion of polymer is ensured by fracturing, and the low water content period is extended.

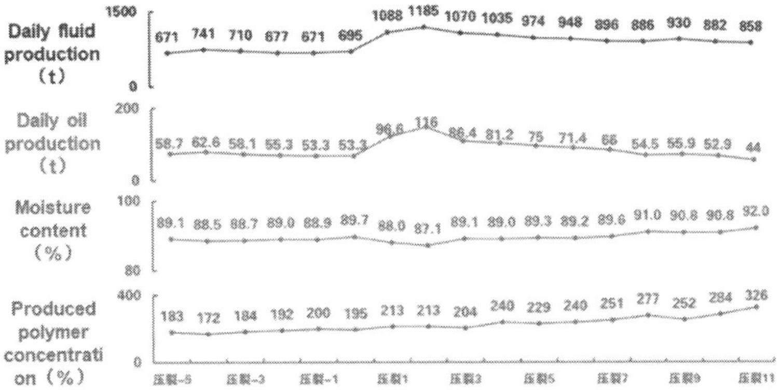

Fig. 3 Drawing curve of fractured well with thick injection and thin production

At the same time, it can be seen from the scatter plot regression curves of different development indicators and unit effective oil increase (FIG. 4) that the relationship between the drop of liquid production before pressure and the drop of water cut before pressure and the increase of oil per unit thickness in the period of low water cut is a monotone increasing exponential function. The greater the drop of liquid production before pressure and the drop of water cut before pressure, the greater the increase of oil per unit thickness, and the better the fracturing effect. The relationship between pre-pressure production and oil gain per unit thickness is monotonically decreasing exponential function. The lower pre-pressure production is, the larger oil gain per unit thickness is, and the better fracturing effect is. Considering the above three types of factors, fracturing can be carried out when the liquid production decline is more than 30%, the water cut decline is more than 5 percentage points, and the production concentration is less than the average value of the whole region, and the daily oil gain of a single well can be more than 3.5t, and the fracturing effect is good.

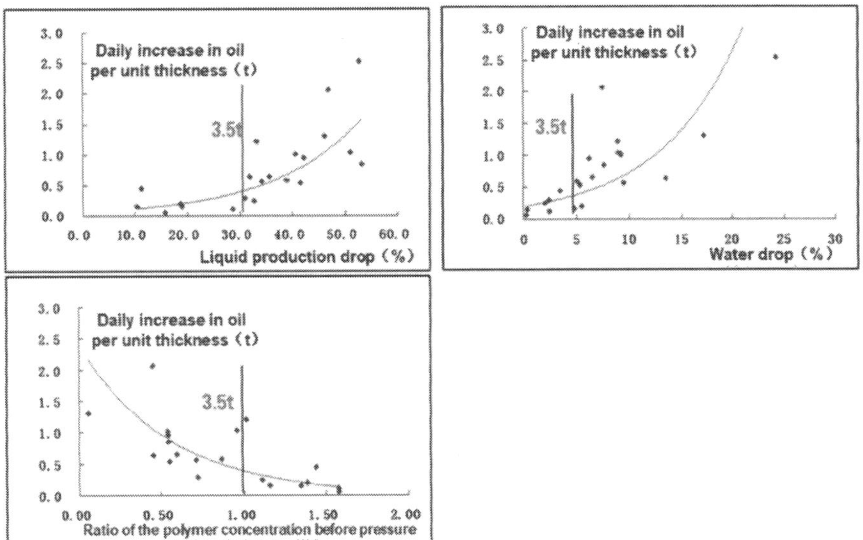

Fig. 4 Scatter plots of different development indicators and unit effective fuel increase

2.3 Thin differential zone Wells with poor production are suitable for fracturing in the water-cut recovery period, and the thin differential zone is the main fracturing object

According to different fracturing layer selection objects, the effects of 65 fracturing Wells in the water-bearing recovery period were classified and analyzed. The results showed that the Wells with thin and differential layers as the main fracturing layer had better effects than those with both good and differential layers. The average daily oil gain per well could reach 3.0t, and the water cut decreased by 3.4 percentage points. However, the decrease in water content is not large. The analysis shows that the Wells with large inter-layer development difference are mainly used in the good zone during the injection process, and the good zone has a higher recovery degree and less potential in the low water cut period. The Wells with both good zone and bad zone are fractured, and the water cut of the whole well is reduced slightly due to the existence of high aquifer for further use. The low aquifer can be exploited by fracturing the thin differential layer, and the water cut of the whole well is reduced.

3. Fracturing the injection well

The proportion of fracturing in blank water drive stage is high. Among 173 fracturing Wells that have been implemented, 94 Wells have been implemented in blank water drive stage, accounting for 54.0% of the total number of fracturing Wells. After putting into operation, these Wells have high injection pressure and poor injection capacity, and in order to meet the conditions of injection convergence, fracturing transformation must be implemented, so the proportion is relatively high. According to the fracturing proportion of blank water drive stage in each block, the fracturing proportion of injection Wells in this stage remains at about 15%.

From the fracturing effects of injection Wells in different stages, it can be seen that the blank water flooding stage has the best fracturing and reconstruction effect, and the effect of pressure reduction and injection increase is obvious. Fracturing Wells in this stage are mainly due to the difficulty of injection caused by oil reservoir pollution during drilling, and their own development and communication are relatively good, so the oil reservoir pollution near the well can be better relieved through fracturing and reconstruction. In the water-cut decline period, the fracturing effect is relatively good, which can achieve the purpose of pressure reduction and injection increase, with an average pressure reduction of 4.3MPa and an average injection increase of 15m^3. The low water cut period and the water cut recovery period are relatively poor, and the effect of pressure reduction and injection increase is not obvious. The average pressure reduction is about 1.7MPa, and the average injection increase is about 7m3. This is mainly due to the fact that with the continuous progress of polymer, the polymer is blocked in the deep oil layer, and the fracturing reform cannot effectively relieve the deep oil layer plugging.

4. Conclusion

(1) The fracturing time of injection well should be dominated by blank water drive stage and water cut down period.

(2) For Wells with low production strength and low flow pressure, the earlier fracturing is performed before 0.5PV injection, the better the effect will be.

(3) The Wells of thick injection and thin production type are suitable for fracturing in the low-water content period, and fracturing is carried out at about 0.2PV, and the final recovery degree is the highest.

(4) The thin and differential Wells with poor production are suitable for fracturing in the water-cut recovery period, which can effectively improve the recovery degree of the well group.

References

1. Bai Jie. A Discussion on fracturing measures of polyflooding extraction Wells in development zone [J]. Journal of Changjiang University (Self Science Edition), 2014,11 (20): 103-104 + 107.
2. Xu Jianjun, Huang Lida, Yan Limei, Yi Na. Insulator Self-Explosion Defect Detection Based on Hierarchical Multi-Task Deep Learning[J]. Transactions of China Electrotechnical Society, 2021, 36(07): 1407-1415.
3. Limei,LIU Yongqiang,XU Jianjun,et al.Broken string diagnosis of composite insulator based on Grabcut segmentation and filler area discrimination [J].Power System Protection and Control, 2021,49(22):114-119
4. Yi, Q. Wang, L. Yan, et al., A multi-stage game model for the false data injection attack from attacker's perspective. Sustainable Energy Grids & Networks 28 (2021).
5. Na Yi, Jianjun Xu,Limei Yan,Lin Huang. Task Optimization and Scheduling of Distributed Cyber-physical System Based on Improved Ant Colony Algorithm. Future Generation Computer Systems, 109(Aug. 2020),134-148.
6. Yang Zhao, Jianjun Xu, Jingchun Wu. A New Method for Bad Data Identification of Oilfield Power System Based on Enhanced Gravitational Search-Fuzzy C-Means Algorithm. IEEE Transactions on Industrial Informatics. VOL. 15, NO. 11, NOVEMBER 2019 5963-5970
7. Jing Han,Xi Wang,LiMei Yan, Aida Dahlak, et al. Modelling the performance of an SOEC by optimization of neural network with MPSO algorithm. International Journal of Hydrogen Energy, Volume 44, Issue 51, 22 October 2019, Pages 27947-27957.https://doi.org/10.1016/j.ijhydene.2019.09.05 5

Practice and understanding of potential tapping technology for ultra short radius horizontal wells

Xiaoyu Zhu*

Daqing Oilfield Co., Ltd The fifth oil production plant Daqing Heilongjiang 163513, China

Abstract. The oilfield has entered the ultra-high water cut period, the distribution of remaining oil is scattered, and it is difficult to tap the potential. On the plane, affected by sand body distribution and well pattern production objects, sand body perforation does not correspond to each other, forming residual oil with imperfect injection production relationship; The residual oil formed in the retention area at the edge of the fault due to the influence of fault shielding. Vertically, affected by the heterogeneity, interlayer and rhythm in the layer, the bottom of the thick oil layer is strongly washed, the middle is moderately strongly washed, and the top is weak and not washed. The production difference is large, and the remaining oil mainly exists at the top of the oil layer. The remaining oil with imperfect injection production relationship and special parts is tapped through ultra short radius horizontal wells to improve oilfield development efficiency.

Keywords: Well layer optimization Trajectory optimization Length of horizontal section Tracking adjustment Effect prediction

1. Well layer optimization

1.1 Well layers with good material basis are preferred

The larger the thickness of sand body, the larger the control area, the larger the reserves, and the better the material basis; The measure effect is calculated based on the limit that the output input ratio is greater than 1. The reserves of the measure layer shall be more than 610t. Considering the above factors, the well layer with a thickness of 1.5m and a reserve of more than 610t is preferred as the target layer.

1.2 Well layers with poor production are preferred

From the perspective of injection production relationship, directional wells can tap the potential of four common remaining oils: imperfect injection production relationship, increasing the effective direction and undeveloped direction of polymer flooding, poor production direction and undeveloped direction of some streamline at the edge of the fault.

1.3 Select well layers with good well conditions

From the perspective of cementing quality, the well condition in the measure well layer and within 20m above is good, especially within the preset window opening range. Good cementing quality is the guarantee for successful window opening.

Through the above analysis, the principle of well and layer selection for ultra short radius horizontal wells is clarified: The remaining recoverable reserves required for the target layer are determined according to the oil price. At present, the remaining recoverable reserves under the oil price of $45 / barrel need to be more than 610 tons; In order to ensure the success rate of sidetracking and the effect of measures, the effective thickness of the target layer is 1.5m and above; The target layer is mainly the residual oil enrichment area near the fault and at the top of the thick oil layer; The cementing quality from the window opening position of sidetracking to the target layer requires good cementation.

2. Well trajectory optimization

2.1 Azimuth and length design

According to the injection production relationship and production status, the design orientation and design track length of the residual oil enrichment area with poor production shall consider the injection production relationship of the well area and the position of the surrounding oil and water wells, and the design length shall be combined with the residual oil reserves on the basis of maintaining a reasonable injection production well spacing.

* Corresponding author: zhxiaoyu@petrochina.com.cn

2.2 Trajectory position design

On the basis of considering the reservoir thickness and reservoir production conditions, in order to ensure the success rate of sidetracking, the track position shall follow the following design. Thick oil layer: for the thick oil layer with effective thickness greater than 2m, the track position is designed at the top of the oil layer; Thin oil layer: for thin oil layers with effective thickness less than 2m and below, the track position is designed in the middle of the oil layer.

2.3 Trend design

According to the reservoir structure, in order to prevent sidetracking deviation or out of formation, the trajectory of horizontal section must be designed along the structural trend, and control points shall be added in areas with horizontal, upward or downward inclination and large structural changes.

Through the above analysis, the trajectory optimization principle of ultra short radius horizontal well is defined: The trajectory orientation should be poor in production and the design of residual oil enrichment area; The length of horizontal section is designed in combination with remaining oil reserves on the basis of maintaining reasonable injection production well spacing; Track position: the thick oil layer is designed at the top of the oil layer, 0.5-1.5m away from the top of the oil layer; The thin oil layer is designed in the middle of the oil layer; The distance between the track and the fault and the end point of the track shall be more than 30m from the adjacent well.

3. Implementation effect analysis

From the overall effect, the average daily oil increase of a single well in the initial stage is 3.2T, and the water cut decreases by 4.1 percentage points. The effect is good. Due to the short stimulation time, the cumulative oil increase in the stage is 1187.9t, and the stage input-output ratio is 1:0.20. In terms of potential tapping types, the best effect is to increase the wells in the direction of polymer flooding effect, with an average daily oil increase of 4.8t in the initial stage, followed by the undeveloped direction of imperfect injection and production, with an average daily oil increase of 4.0t in the initial stage, followed by the potential tapping effect at the edge of the fault, with an average daily oil increase of about 2.0T in the initial stage. In order to implement synchronous adjustment to the surrounding oil and water wells after extending the measures, first, connect the water wells to improve the liquid supply capacity; The second is to improve the utilization degree of injection layer; Third, the well shall be exploited at a reasonable speed; Fourth, the plane of adjacent oil wells shall be adjusted accordingly. Through the corresponding adjustment of both ends of oil and water wells, the injection production structure is optimized, the injection production balance is maintained, and the effect of protection measures is improved. At the same time, the production of 8 sidetracking horizontal wells is regressed by regression method, and the decline law of sidetracking horizontal wells is obtained, which conforms to exponential decline. It is predicted that the cumulative oil increase of 8 measures is 10212t, which is calculated at USD 45 / barrel, and the input-output ratio is 1:1.68.

4 Prototype basin thickness recovery

It can be seen from Fig.1 that the original strata of the Middle-Upper Triassic are mainly distributed in Taibei sag, Tainan sag, Keyayi sag, Toksun sag, Buerjia uplift, Luxi uplift and Takequan uplift. The original thickness is 200m-1200m. The sedimentary center is located in the front of Bogda Mountain and Shaquan 2 well. The sedimentary thickness of Bogda Mountain is 1200 m, and the sedimentary thickness of Shaquan 2 well is 1000 m. The thickness of the original strata in the western Tuha Basin gradually increases from south to north, reaching the maximum in the front of the Bogda Mountains.

In the Zhundong area, the thickness of the original strata in the southern part of Beisantai uplift and Shaqi uplift is thickened from north to south, and the sedimentary thickness is 0m-600m. In the Jimsar sag, the thickness of the original strata is thickened northward, and the thickest is 400m (Figure 1).

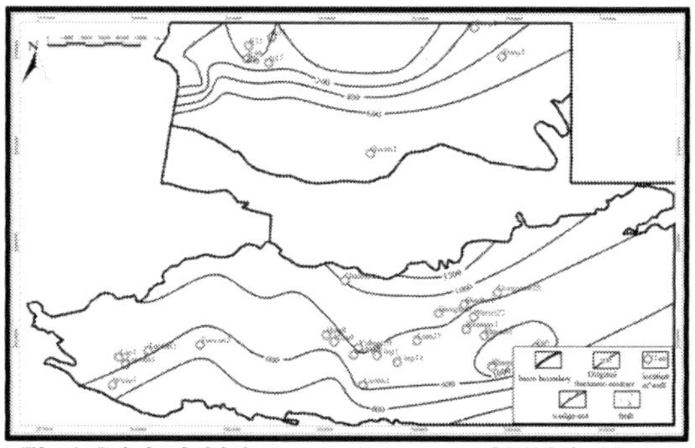

Fig.1 Original thickness contour map of Middle-Upper Triassic in western Turpan-Hami Basin-eastern Junggar Basin

5 Prototype basin framework

The results were calculated by denudation. Combined with the residual formation thickness of each layer, the original formation thickness of each layer can be obtained by calculation. On this basis, combined with the current tectonic characteristics and tectonic evolution process, the prototype basin is restored, and compared with the current structure, the ancient tectonic pattern and basin prototype evolution can be further analyzed.

In the Middle-Upper Triassic sedimentary period, Bogda Mountain belongs to the sedimentary product of stable depression type, and Bogda Mountain and its adjacent areas show a quasi-plainization process of cutting high and filling low [4]. The western part of the Tuha Basin is a depression-type basin. The Taibei and Tainan depressions are further expanded, the sedimentary center moves southward, the Pubei low uplift becomes an erosion area, and the Keyayi and Tuokexun depressions are developed on the west side. The tectonic framework

https://doi.org/10.1051/e3sconf/202447801019

of the Triassic in the eastern Junggar Basin is consistent with that of the Upper Permian. The Beisantai uplift and Shaqi uplift in the north continue to develop and expand. The southern side is still the Zhundong depression area (Figure 2).

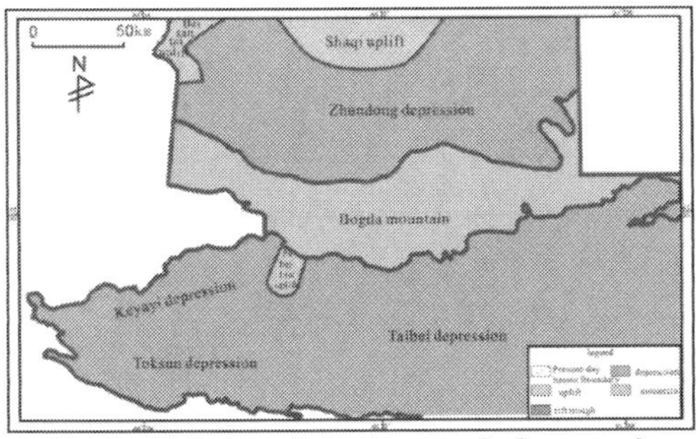

Fig.2 Middle-Late Triassic tectonic framework

6 Strata erosion thickness recovery

In this study, the erosion amount of the middle and upper Triassic on the profile was calculated according to the seismic stratigraphic trend thickness method, and a plane distribution map of erosion amount was formed (Fig.3).

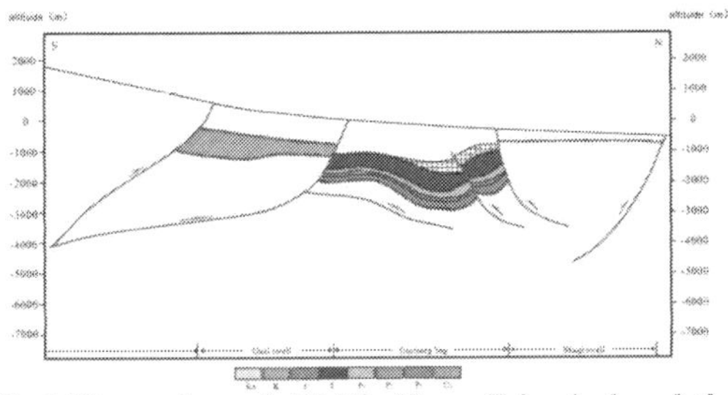

Fig.3 The north-south Middle-Upper Triassic denudation interpretation profile in the east of Zhundong area

It can be seen from Fig.3 that the Indosinian movement at the end of the Late Triassic made the western part of the Turpan-Hami Basin-the eastern part of the Junggar Basin once again subjected to near north-south compression, and the strata in the northern and southern margins of the basin were uplifted and eroded. In the western region of Tuha Basin, the erosion range of the Middle-Upper Triassic strata is larger than that of the Upper Permian, which is distributed in Taibei Sag, Tainan Sag, Keyayi Sag, Toksun Sag, Buerjia Uplift, Luxi Uplift and Takquan Uplift. The erosion thickness is 0m-400m ; the erosion thickness on the east side of the western Tuha Basin gradually increased from Dongtan 1 well to Dongshen 2S to the north to 400 m, and the erosion thickness to the south increased first and then decreased, reaching the maximum at the slope and decreasing southward .
In this study, three seismic profiles were selected in the western and Zhundong areas of the Tuha Basin to track the unconformity, identify and compare the erosion points and erosion characteristics of the unconformity interface (Figure 2). Based on the above principles, the erosion amount of the Upper Triassic in each section is calculated on the basis of 6 seismic sections (Fig.3).

7. Conclusion and summary

(1)Changing thinking is the premise. In the late stage of ultra-high water cut, the distribution of remaining oil is more and more scattered, and it is more and more difficult to tap the potential conventionally, which seriously restricts the development effect of the oilfield. We should face up to the development contradiction between effect and benefit, change our ideas, dare to explore and practice, and explore an effective development mode integrating cost, oil increase and benefit.
(2)Potential enrichment is the basis. In order to ensure the oil increase effect of sidetracking, the well layers with large thickness and large reserves that are not developed or poorly developed should be selected for layer selection. At the same time, combined with the distribution of residual oil saturation, it should ensure that the target layer of the measure has rich residual potential.
(3)Trajectory design is the key. The reservoir is thin and thick, the structure is high and low, and the producing parts are up and down. The accurate trajectory parameter design is the key to ensure the sidetracking effect.
(4)Protection measures are the guarantee. Optimize injection production adjustment, implement protection measures, provide sufficient energy supply, determine reasonable injection production intensity, prolong the effect of measures and ensure the maximum effect.

References

1. Li Yang. 2001. Heterogeneous model of channel sand reservoir. Science Press.
2. Xu Jianjun, Huang Lida, Yan Limei, Yi Na. Insulator Self-Explosion Defect Detection Based on Hierarchical Multi-Task Deep Learning[J]. Transactions of China Electrotechnical Society, 2021,36(07):1407-1415.
3. Limei, LIU Yongqiang, XU Jianjun, et al. Broken string diagnosis of composite insulator based on Grabcut segmentation and filler area discrimination [J]. Power System Protection and Control, 2021, 49(22): 114-119
4. Yi, Q. Wang, L. Yan, et al., A multi-stage game model for the false data injection attack from attacker' s perspective. Sustainable Energy Grids & Networks 28 (2021).
5. Na Yi,Jianjun Xu,Limei Yan,Lin Huang. Task Optimization and Scheduling of Distributed Cyber-physical System Based on Improved Ant Colony Algorithm. Future Generation Computer Systems, 109(Aug. 2020),134-148.
6. Yang Zhao, Jianjun Xu, Jingchun Wu. A New Method for Bad Data Identification of Oilfield Power System Based on Enhanced Gravitational Search-Fuzzy C-Means Algorithm. IEEE Transactions on

Industrial Informatics. VOL. 15, NO. 11, NOVEMBER 2019 5963-5970

7. Jing Han,Xi Wang,LiMei Yan, Aida Dahlak, et al. Modelling the performance of an SOEC by optimization of neural network with MPSO algorithm. International Journal of Hydrogen Energy, Volume 44, Issue 51, 22 October 2019, Pages 27947-27957.https://doi.org/10.1016/j.ijhydene.2019.09.055

Smart Transformation Technology for Digitizing T Oilfield

Shuxin Sun

10th Oil Extraction Plant, Daqing Oilfield Limited Liability Company, Daqing City, Heilongjiang Province, 163000, China

Abstract: In response to the current situation of T Oilfield being remote, with high total labor demand, dispersed well distribution, and low automation levels, this paper systematically summarizes the digital technologies and application effects in the construction of S operation area. By integrating process optimization, streamlining management levels, and incorporating video and communication technologies, a low-cost digital construction model for T Oilfield is established. During the "14th Five-Year Plan" period, with the goal of reducing construction investment and streamlining management levels, the low-cost digital construction model is applied to adjust the organizational structure of T Oilfield, reduce total labor demand, and achieve cost-effective, digitally lean management.

Keywords: Unmanned operation, Centralized monitoring, Optimization, Operation and maintenance.

1. Overview

The approach to the digitization construction of T Oilfield involves leveraging mature station control construction technology and integrating it into a centralized monitoring model. In the design process of constructing the ST Joint Station, an unmanned design is adopted for each production unit, with minimal personnel in the comprehensive control room implementing monitoring. In this capacity expansion construction, newly built oil-water wells and oil collection and distribution stations are digitally constructed according to the standards of digital construction. The newly constructed ST Joint Station is designed with a centralized monitoring and unmanned operation mode, and the collected data is wirelessly transmitted to the newly built production management center. The centralized control room in the ST Joint Station conducts centralized monitoring. Video surveillance is installed at platform well sites, and the video signals are transmitted wirelessly.

2. Design Concept

To meet production needs, ensure safe and reliable operation, and enhance production management, a PLC control system is employed for centralized monitoring and control of process parameters. The operating station serves as the human-machine interface, utilizing on-site instruments for automated detection and control of the production process. All parameters are fed into the control system for display, alarms, interlocking, recording, and printing. Signals from individual instruments within the station are connected to remote I/O cabinets at the operator's station. Control unit signals are then accessed by the central control room through the station's internal network.

2.1 Establishing the Central Control Room

In this project, a central control room is established in the comprehensive control room, mainly comprising the control room and equipment cabinet room. The central control station is constructed according to the standards for important facilities and electronic computer rooms throughout the entire station. The operation stations and centralized display stations are set up based on the number of acquired I/O points.

2.2 Optimize Internal Processes at the Station

The heating process adopts a phase-change vacuum heating furnace, with the water treatment process for heating water and the pumping method retained. The heating equipment is installed outdoors. The technical advantage of the phase-change heating furnace is that there is no risk of expansion, explosion, or scaling, making it safe and reliable. The equipment is compact, equipped with its own control device, making it convenient for control. It operates without the need for constant supervision, providing more convenient management.

2.3 Remote Control

Equipment, whether self-contained or third-party control devices, should have open communication protocols, including but not limited to: RS485: MODBUS RTU or PROFIBUS DP; MODBUS TCP/IP or DNP3. The specific project's design technical documents should specify the communication method, protocol standards, and version adopted. Third-party control equipment suppliers should provide a data interface condition table, etc., for the PLC to obtain data in a read-only manner.

When variable frequency drives (VFDs) are used for rotating process equipment, the following signals need to be transmitted to the PLC through hardwiring or communication: frequency control, frequency, equipment power frequency/VFD operation status, VFD operation status, VFD fault alarms, and VFD start. If the original equipment does not have this functionality, cooperation with the manufacturer is required to install it on the original equipment.

2.4 Installation of Video Security System

To establish safety and security for personnel and equipment, the station is equipped with video surveillance and alarms, with a focus on preventing intrusion and monitoring for any signs of leakage.

The entire station is equipped with a network-based video surveillance system, with a total of ** sets of surveillance front-end installations. The dual three joint station's video monitoring images are connected to the comprehensive control room via optical fiber. A new industrial network switch is installed in the room to manage and call up video monitoring images within the jurisdiction, and to store video monitoring images in real-time. In the control room, * sets of new LCD monitors are installed, displaying well and interstitial video images, site video images, video monitoring management images, and perimeter intrusion alarm images. The backend equipment of the perimeter intrusion alarm system is installed in the comprehensive control room's equipment cabinet room and is linked with the industrial television monitoring system. Management personnel can confirm on-site situations based on the monitoring images. The central control uses PLC + remote I/O, and the * production units within the station operate unmanned. Since 2020, the station's production operation has been centrally monitored, and it is currently running smoothly.

3. Unmanned Transformation of Oil Collection and Water Distribution Unit

1) Process instruments adopt wireless communication, while flow instruments use a bus system, making construction simple and valve groups aesthetically pleasing.

2) The control unit uses RTU, leveraging its high integration, small size, easy layout, and support for wireless communication.

3) Data transmission is carried out wirelessly, requiring less investment, facilitating networking, avoiding the difficulties of wired network installation, and addressing the issue of high investment.

This technological approach is implemented in the construction of new valve groups, using a mesh network approach, which has proven effective. It has successfully achieved the unmanned management of ** oil collection and water distribution valve groups and * oil transfer stations. Subsequently, it has been widely promoted in areas such as T.

4. Pumping Unit and Water Injection Well Transformation

Exploring the technological direction of digitizing oil-water wells, the shift to outdoor work environments necessitates comprehensive consideration of instrument explosion-proofing, low-temperature resistance, and data transmission modes. After technical investigations and comparisons, the following approaches are adopted:

1) Pumping wells have an integrated wireless condition acquisition unit for load, displacement, and stroke collection at the wellhead, as well as a wellhead pressure detection unit and current and voltage detection devices.

2) Single-tube single-well injection wells are equipped with a high-pressure flow detection and control device, along with inlet and outlet pressure detection, flow measurement, and regulation. Power can be sourced from the nearby pumping unit platform transformer. In cases where nearby power sources are unavailable, solar power generation devices can be installed.

3) The wellsite's integrated control device, considering comprehensive investment and maintenance needs, uses DTU technology for wireless radio parameter analysis controllers (single well, main well) and intelligent wireless radio parameter analysis control modules (for secondary wells). It employs more compact and intelligent data acquisition technology.

4) Considering the decentralized nature of T Oilfield well sites and the scarcity of mobile base stations in the area, wireless transmission adopts the method of establishing self-built wireless base stations.

After preliminary research and in accordance with relevant national regulations, available technology solutions for meeting the requirements of transmitting video bandwidth over a wide area include self-built LTE, wireless bridges, leasing operator 4G networks, WIAFA technology, and ADSS fiber optic cables.

In compliance with network security regulations, the use of public network 4G for transmitting production data must go through a demilitarized zone (DMZ) for isolation before connecting to the office network. Due to the long transmission path from the data collection equipment to the monitoring terminal, there is a high failure rate, and the number of DMZ points is limited, making expansion difficult. The construction cost of ADSS fiber optic cables is approximately 35,000 RMB/km, and it is costly to deploy them for all wells and sites outside the station. Wireless bridges and WF technology require a line of sight with no obstructions for transmission between two points.

5. Construction of the Technical Application for the Production Management Center

Utilizing IoT (Internet of Things) technology, a system is established to centrally manage and control production data and equipment status information from oil and gas field well areas, measurement stations, oil transfer stations, and joint stations at the production command center and production control center of the oil and gas production plant. The system comprises data collection and monitoring subsystems, data transmission subsystems, and production management subsystems, collectively forming a large-scale SCADA (Supervisory Control and Data Acquisition) system. It constitutes an information management system that incorporates real-time monitoring of production data, production analysis, safety warnings, operational scheduling, and data management, all built using data processing and data analysis technologies.

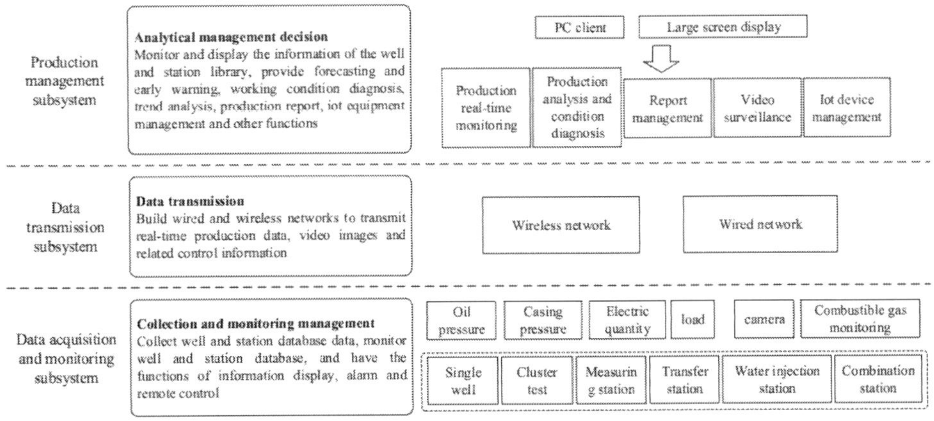

Fig.1 Overall system framework

In the construction of this system, the following main technologies are applied:

(1) Application of Database Technology:

The relational databases of the oil and gas field company's production management subsystem should be built using the databases specified by the headquarters. The dynagraph relational database in the operational area should also be constructed using the databases specified by the headquarters. The database should consider technical reliability, openness, and scalability. The real-time database should support multiple servers and server clusters, as well as multiple CPUs. It should also support a distributed structure and distributed data collection and processing.

The real-time database should adopt a modular structure, making it easy to add new application modules to the real-time database software. Hardware devices should be designed, deployed, and upgraded based on the data volume and user access at each deployment point.

(2) Application of Network Technology:

As the construction of the production IoT progresses, distinct from the internet, the IoT has strict requirements for professionalism and security. Considering the undulating terrain, complex topography, and the area being a flood-prone zone along the river between the ST Joint Station and the production management center, this project uses point-to-point long-distance bridge equipment to achieve data upload.

The straight-line distance between the ST Joint Station and the production management center in the operational area is approximately ** km. The data transmission is achieved through point-to-point wireless communication using 45m communication towers at the Oilfield production management command center and 35m communication towers (constructed simultaneously) to ensure line-of-sight wireless transmission.

(1) Data Transmission:

The production management center receives production and office network data transmitted by the wireless communication system, which is then laid directly through buried optical cables to the equipment cabinet room at the Oilfield production management center. One set of Layer * industrial switches is installed, with two sets of Layer * industrial switches underneath. One is used for transmitting production data, and the other is used for transmitting video data.

(2) Isolation of Production Network and Office Network:

The isolation of the production network and office network uses firewalls and unidirectional isolation gateways, allowing only one-way transmission of production network data to the office network to ensure the security of the production network.

6. Analysis of Application Effects

After applying this construction model, three transformations are achieved: (1) a shift from manual operations to fully automated processing, (2) a transition from intermittent data points to continuous information flow, and (3) a shift from experience management to refined management.

6.1 Application Effects in Implemented Regions

Through the construction in the region, a two-tier digital monitoring architecture has been established, consisting of the work area and operational area. An integrated operational area-level production management center has been created, incorporating regional production anomaly alarms, statistical summaries, and decision-making commands. In the region, data collection for * joint station, * oil transfer stations, and ** measurement stations has transitioned from manual to automatic, with reports generated automatically. The management mode has shifted from manual periodic inspections to inter-station video inspections, optimizing frontline production staffing by nearly ** people. This has successfully achieved the established goals of centralized monitoring for joint stations and oil transfer stations, as well as unmanned operation for measurement stations and water distribution stations.

6.2 Predicted Application Effects After Promotion

Based on the digital production management model and considering geographical proximity, adjustments and mergers of mining-level production units are made. The entire plant is divided into * production blocks, with each block having one regional production management center, centered around large-scale station areas. Following the transformation, it is expected to dissolve 4 mining-level production units, reducing the total frontline production staff from 2*** people to 1*** people, a decrease of 1*** employees, representing a reduction of 47.5%. This is estimated to result in an annual savings of 158 million yuan in labor resources costs.

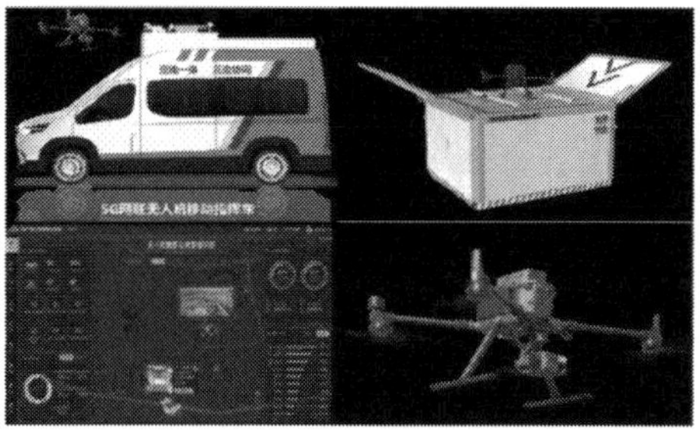

Fig.2 Overall Intelligent and unmanned monitoring methods

7. Conclusion

Currently, frontline operational staff is aging, and the availability of labor is becoming increasingly tight. The challenging situation demands the optimization and innovation of traditional management methods. Innovation in management is inseparable from the innovation and use of new technologies. It requires a continuous absorption of advanced design concepts and innovative application of technology. In future designs, the introduction of advanced technology will replace manual daily inspections and operations. Simultaneously, the reduction in operational staff requires higher equipment maintenance demands and greater needs. Therefore, it is imperative to establish a comprehensive maintenance system, intensify efforts in cultivating digital professionals, and effectively ensure the smooth operation of unmanned guarding systems on-site.

References

1. Jing Han,Xi Wang,LiMei Yan, Aida Dahlak, et al. Modelling the performance of an SOEC by optimization of neural network with MPSO algorithm. International Journal of Hydrogen Energy, Volume 44, Issue 51, 22 October 2019, Pages 27947-27957https://doi.org/10.1016/j.ijhydene.2019.09.055
2. Xu Jianjun, Huang Lida, Yan Limei, Yi Na. Insulator Self-Explosion Defect Detection Based on Hierarchical Multi-Task Deep Learning[J]. Transactions of China Electrotechnical Society, 2021,36(07):1407-1415.
3. Limei,LIU Yongqiang,XU Jianjun,et al.Broken string diagnosis of composite insulator based on Grabcut segmentation and filler area discrimination [J].Power System Protection and Control, 2021, 49(22):114-119
4. Yi, Q. Wang, L. Yan, et al., A multi-stage game model for the false data injection attack from attacker's perspective. Sustainable Energy Grids & Networks 28 (2021).
5. Na Yi,Jianjun Xu,Limei Yan,Lin Huang. Task Optimization and Scheduling of Distributed Cyber-physical System Based on Improved Ant Colony Algorithm. Future Generation Computer Systems, 109(Aug. 2020),134-148.
6. Yang Zhao, Jianjun Xu, Jingchun Wu. A New Method for Bad Data Identification of Oilfield Power System Based on Enhanced Gravitational Search-Fuzzy C-Means Algorithm. IEEE Transactions on Industrial Informatics. VOL. 15, NO. 11, NOVEMBER 2019 5963-5970

Experimental study on the effect of confining pressure and interval time on the fatigue mechanical properties of mudstone

Yuhao Zhang [1,*], Lachun Ren [1], Peng Yao [1], Deyi Jiang [2,3]

[1] Powerchina Urban Planning and Design Institute Co. Ltd., Guangdong 511400, China
[2] School of Resources and Safety Engineering, Chongqing 400044, China
[3] State Key laboratory of coal mine disaster dynamic and control, Chongqing 400044, China

Abstract. The mudstone interlayer experiences discontinuous cyclic loading during gas storage reservoir operation. To investigate the effects of intermittent action and confining pressure on the fatigue mechanical properties of mudstone. Continuous fatigue test and graded intermittent fatigue test were conducted respectively. The results show that: 1. the fatigue life of mudstone is enhanced by the intermittent effect during the fatigue process. The presence of confining pressure under the same experimental conditions also benefits the fatigue life of mudstone.2. the average modulus of elasticity of mudstone decreases with increasing stress level during cyclic loading, and both intermittency and confining pressure help to enhance the modulus of elasticity of mudstone.3. in graded intermittent fatigue tests. In both uniaxial and triaxial tests, the residual strains generated by the post-interval cycles are smaller than those generated before the intervals.4. In the 3MPa graded interval fatigue test, negative deformation occurs in the post-interval cycle due to the release of elastic energy.

1. Introduction

Due to the need for energy security reserves, many underground salt cavern gas storage facilities have been built in China in recent years. During the operation of underground gas storage, with the injection and extraction of gas. The surrounding rock experiences discontinuous cyclic loading action. As shown in Fig.1 for Germany huntorf power station daily peak pressure diagram. Unlike foreign salt rock is distributed in the form of giant salt mounds in the underground, salt rock in China is mostly distributed in the form of layered salt rock. The salt rocks contain interlayers such as mudstones. Mudstone is a weakly consolidated clay that has formed a strongly consolidated rock after a moderate degree of post-generation [1], In general, its uniaxial compressive strength is low, belonging to the soft and weak rock, which is extremely widely distributed, and shale accounts for about 50% of the total amount of the surface [2].

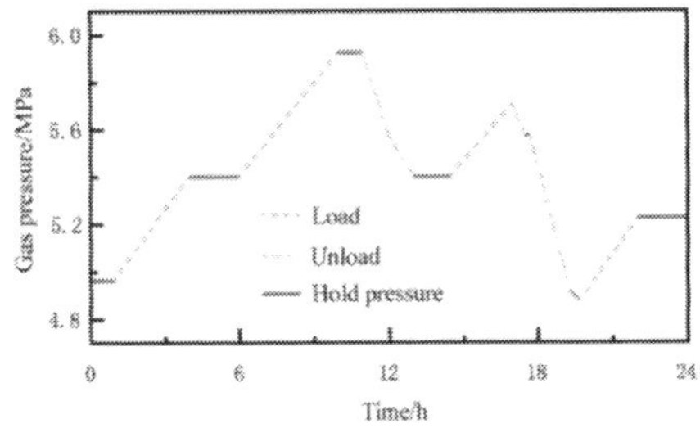

Fig.1 Diagram of a day's operating pressure of a pressurized gas storage plant

In order to ensure the safe operation of underground gas storage reservoirs. Considering the underground storage reservoir surrounding rock in the storage reservoir operation process by the load is often in the change, a large number of scholars so on the rock grading load under the mechanical properties of research [3-7]. Chen Wenwu et al.[8] conducted graded loading test with red layer soft rock and found that the theoretical curve of Burgers model is basically consistent with the test curve, which can better describe the creep characteristics of red layer soft rock; Lu Gao Ming et al.[9] found that the fatigue life of yellow sandstone can be predicted by the number of cycles-upper limit stress curve under a certain condition; Xu Ying et al. [10] investigated the energy evolution and damage characteristics of mudstone under cyclic loading and unloading conditions; Cai Guojun et al [11] investigated the kinetic properties of mudstone under cyclic loading and unloading and found that there was a quadratic polynomial relationship between the damage variables and the strain of the mudstone; Guo Yindong et al [12] carried out fatigue tests under uniaxial cyclic loading of mudstone, studied the fatigue strength, deformation, and damage characteristics of mudstone, and made an analysis of the evolution of fatigue damage of mudstone under cyclic loading and the fatigue strength, deformation, and damage characteristics of mudstone under uniaxial cyclic loading. Under the action of cyclic loading, the fatigue damage evolution law of salt rock was initially explored; some scholars conducted experiments to investigate the effect of fatigue characteristics of rocks under the action of time interval, Li Zongze et al [13]

* Corresponding author: 1830329412@qq.com

found that the residual strain produced by cycling after the time interval was greater than that before the interval; Jiang Deyi et al [14] found that the intermittent fatigue life of salt rock was shortened more than 50% compared to the traditional fatigue life; Fan et al [15] found that with the interval time, the fatigue strength, deformation and damage characteristics of salt rock were investigated. found that the fatigue life of salt rock decreases and the residual strain increases with the increase of interval time; related studies show that factors such as temperature and humidity have an effect on the mechanical properties of rocks under cyclic loading and unloading [16 17]; Wang Ruihong et al [18] investigated the relationship between residual strength of the rock body and cyclic loading, and found that there is a negative correlation between the two; Yang Yongjie et al [19] found that the coal rock is more prone to fatigue compared to Zhu Zhuohui et al [20] found that the overall strength of the specimen decreased under uniaxial cyclic loading; He Mingming et al [21] established a damping ratio relationship model that can describe the damping behaviour during graded cyclic loading; Feng Chunlin et al [22] found that the upper limit stress and amplitude of the cyclic loading are the main factors affecting the fatigue damage process and fatigue life of white sandstone; Meanwhile, a large number of scholars have studied the relationship between residual strength of mudstone and cyclic loading in mudstone. Meanwhile, a large number of scholars have also carried out experimental investigations on the mechanical properties of mudstone under cyclic loading and unloading [23-30].

Existing studies have many results on the fatigue creep mechanical properties of rocks. There are fewer studies on the fatigue mechanical properties of mudstone. Considering the existence of discontinuous loading in the actual operation of the gas storage reservoir, and the special layered salt rock structure in China. The weak mudstone interlayer should be studied. In this paper, by designing graded uniaxial continuous fatigue test, graded uniaxial intermittent fatigue test and graded triaxial intermittent fatigue test. It explores the influence of the presence of confining pressure and intermittent time on the mechanical properties of mudstone, which is of great significance to ensure the safe operation of the gas storage reservoir.

2. Test conditions and program

2.1 Test apparatus and specimen preparation

The mudstone used for the tests was from Changsha City, Hunan Province and had a yellowish-brown appearance. According to the standard of International Society for Rock Mechanics (ISRM), the specimens were processed into standard cylindrical specimens with a height of 100 mm and a diameter of 50 mm (height to diameter ratio of 2:1), and in order to reduce the influence of the unevenness of the end surfaces on the test results, the top and bottom end surfaces were polished so that the unevenness of the specimens was controlled to be less than ±2 mm. Considering that the difference in water content of mudstone has a great influence on the test results, the specimens were stored in a drying box immediately after the specimens were processed to keep them dry. The specimens are shown in Fig. 1(a).

The test was carried out in the State Key Laboratory of Coal Mine Hazard Dynamics and Control of Chongqing University, and the instrument used was the MTS815 Rock Mechanics Test System produced by MTS Corporation of the United States of America, which is mainly used to test the mechanical properties and seepage characteristics of rock, concrete, and coal under complex stress conditions, such as splitting, uniaxial compression, triaxial compression, and cyclic loading and unloading tests. The testing accuracy is high performance and stable, and can be used for high and low speed data acquisition, with force, displacement axial strain, lateral strain and other control modes. The equipment axial maximum loading load up to 2800KN, confining pressure up to 80MPa, pore water pressure up to 80MPa, temperature up to 200 °C. The test equipment is shown in Fig. 1(b).

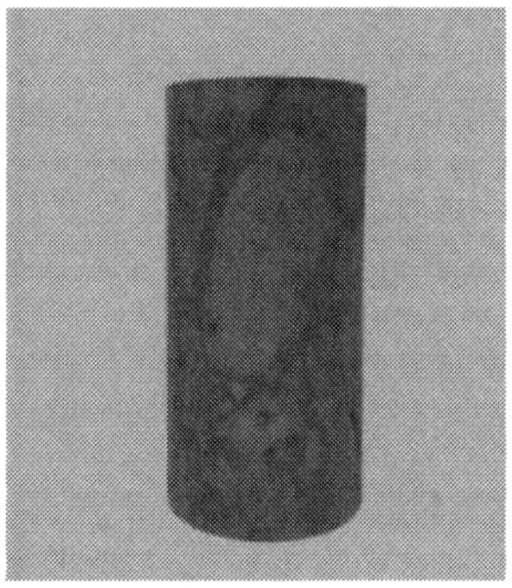

(a)

(b)

Fig.2 The manufactured mudstone sample, MTS815 rock mechanics testing system.

2.2 Test program

The mudstone specimens were first subjected to uniaxial compressive strength testing versus triaxial compressive strength testing with a confining pressure of 3 MPa. The test data are shown in Table I.

https://doi.org/10.1051/e3sconf/202447801021

Table1 Compressive strength of mudstone

Confining pressure/ MPa	Specimen number	Compressive Strength /MPa	Average compressive strength /MPa
0	0-1	12.01	12.37
	0-2	12.91	
	0-3	12.19	
3	3-1	31.5	32.67
	3-2	33.7	
	3-3	32.8	

The test program is shown in Fig. 4:

Fig. 2(a) shows a continuous fatigue test with a test load rate of 0.1 kN/s, setting the stress level at the first level of 30% of the mean uniaxial compressive strength. The step-by-step increment is 20%. The lower limit of stress is 10% of the average compressive strength. Each level of cyclic loading and unloading 10 times.

Fig. 2(b) for the graded upper limit interval fatigue test(GUIF), the test load rate of 0.1kN / s, set the stress level of the first level for the average uniaxial compressive strength of 30%. The step-by-step increment is 20%. The lower stress limit is 10% of the average compressive strength. Each level of cyclic loading and unloading 10 times, every two cycles in the upper stress limit constant 4 minutes. A total of 2 sets of tests were conducted with a confining pressure of 0 MPa and a confining pressure of 3 MPa.

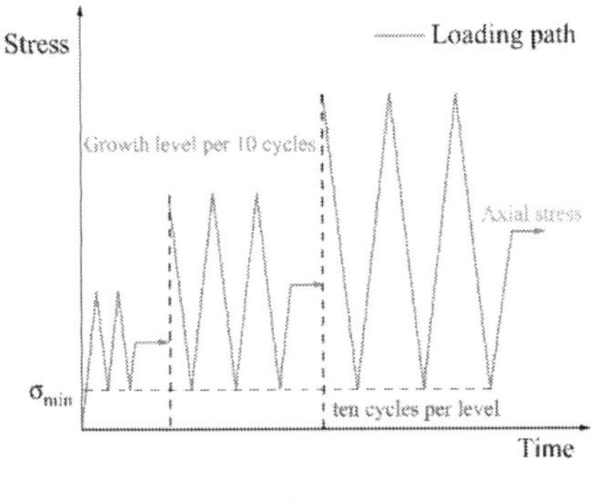

(a)

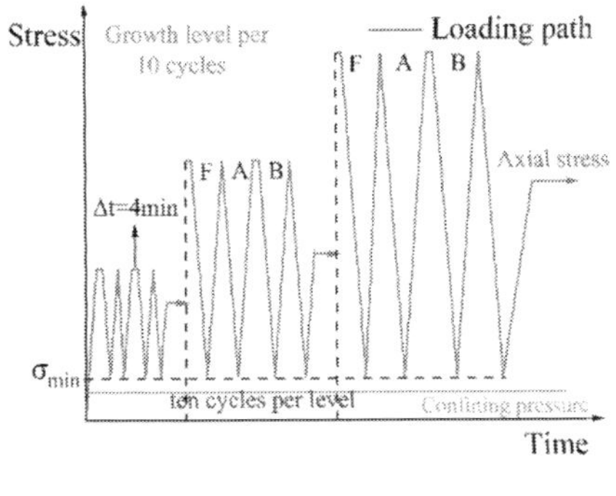

(b)

Fig.3 Schematic diagram of test scheme

3. Test results and analyses

3.1 Axial Strain Change Laws

As shown in Fig. 3(a) for the continuous fatigue test stress-strain curve, Fig. 3(b) for the uniaxial intermittent fatigue test stress-strain curve, and Fig. 3(c) for the triaxial intermittent fatigue test stress-strain curve.

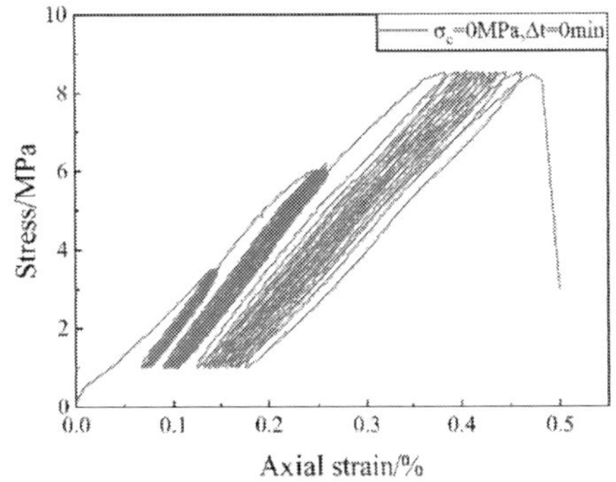

(a)

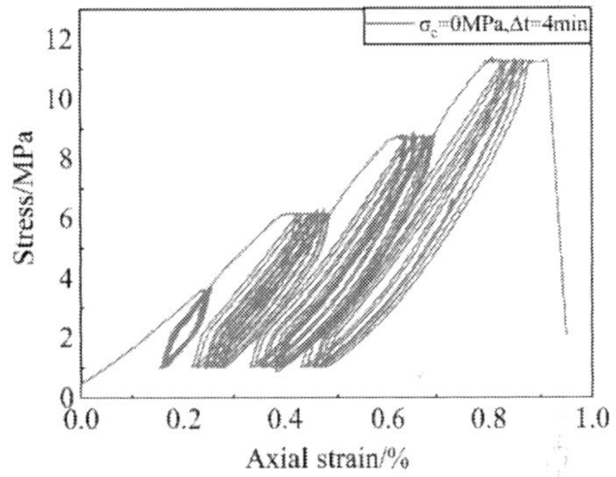

(b)

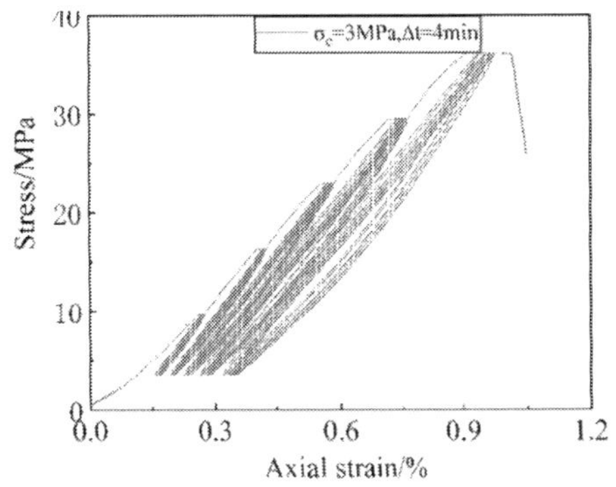

(c)

Fig.4 Stress-strain curve of test results(a)σ_c=0MPa,Δt=0min, (b)σ_c=0MPa,Δt=4min, (c)σ_c=3MPa,Δt=4min

In this paper, the number of cycles experienced by the mudstone before destruction is taken as the fatigue life. Comparing Fig. 4 (a) and (b) it can be easily seen that the intermittent time enhances the fatigue life of mudstone.

The fatigue life of the continuously graded uniaxial fatigue test is 30, and the fatigue life of the graded uniaxial intermittent fatigue test is 40.Comparing Fig. 4(b) and (c), it can be seen that the presence of confining pressure also enhances the fatigue life of the mudstone. The fatigue life of 3 MPa graded triaxial intermittent fatigue test mudstone is 50 under the same test programme as graded uniaxial intermittent fatigue test.

The reason for the intermittent time to enhance the fatigue life of mudstone is. During the graded cyclic loading process, during the period of constant upper stress limit. The mudstone is loaded internal particles squeeze each other to fill, the original pore cleavage is closed under this action. The mudstone becomes denser, which improves its ability to resist damage and deformation. The ultimate load it can withstand is also greater, so the fatigue life is increased.

And the reason for the enhanced fatigue life under confining pressure is. Compared with the uniaxial test, in the triaxial test. Under axial loading, the presence of confining pressure restricts the deformation expansion of mudstone in radial direction, and under this extrusion, it causes the closure of the primary fissure pores of mudstone, and the cracks that expand during cycling also partially undergo closure under the action of confining pressure. Therefore, the fatigue life of mudstone is also enhanced by confining pressure.

3.2 Elastic modulus

The elastic modulus, also known as Young's modulus, is one of the most important and characteristic mechanical properties of rock materials and characterizes the degree of difficulty of elastic deformation of an object. During the loading process, mudstone will produce elastic and plastic deformation, while only elastic deformation can be recovered during the unloading process. In order to reflect more intuitively the influence of different test factors on mudstone deformation, we use the modulus of elasticity of mudstone during unloading as the modulus of elasticity E in this paper, which reflects the ability of mudstone to recover from elastic deformation. Its calculation formula is as follows:

$$E = \Delta\sigma / \Delta\varepsilon \tag{1}$$

where $\Delta\sigma$ is the stress amplitude and $\Delta\varepsilon$ is the strain difference.

In order to avoid the influence of cyclic loading and unloading on the modulus of elasticity under the same grade, the modulus of elasticity of each cyclic unloading stage was counted and averaged as the modulus of elasticity of mudstone under the grade to reduce the error. As shown in Fig. 5, the variation of the average modulus of elasticity of mudstone specimens with the stress level under different test schemes.

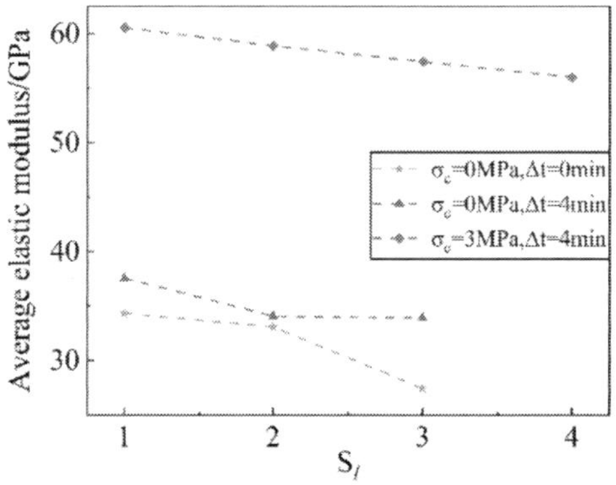

Fig.5 Variation of Elastic Modulus with Stress Level

It can be found by observing Fig. 5. Whether in uniaxial continuous fatigue test, graded uniaxial fatigue test or graded triaxial fatigue test, the average modulus of elasticity of mudstone decreases gradually with the increase of stress level. This is because as the stress level increases, the increase in stress amplitude leads to greater damage to the mudstone by the loading action, and the modulus of elasticity characterizes the ability of the rock to resist deformation damage. The modulus of elasticity is a characterization of the ability of rocks to resist deformation and damage.

The average modulus of elasticity in the triaxial test is significantly higher than that in the uniaxial test. This is because the presence of confining pressure in the triaxial test restricts the expansion of radial deformation of mudstone, which enhances the ability of mudstone to resist deformation damage, and therefore the modulus of elasticity of the triaxial test is larger than that of the uniaxial test under the same amplitude ratio.

3.3 Residual strain

The deformation of rock produced during loading consists of a combination of recoverable characteristic strain and unrecoverable residual strain of deformation.

$$\varepsilon = \varepsilon_e + \varepsilon_r \tag{2}$$

Where the upper stress limit corresponds to the strain that is ε , the lower stress limit corresponds to the strain that is ε_r , ε_e is the elastic strain.

The initial cycle of each stress level is the first cycle, then the cycle before the interval period in the loading process is recorded as the A cycle, and the cycle after the interval period is recorded as the B cycle, and the residual strains generated in the process of A and B cycles are counted respectively, as shown in Fig. 3.

Fig. 6(a) shows the residual strain of uniaxial graded upper limit intermittent fatigue test, and Fig. 6(b) shows the residual strain of 3MPa graded upper limit intermittent fatigue test.

Both uniaxial intermittent fatigue test and triaxial intermittent fatigue test. The residual strain before the interval is greater than the residual strain after the interval. This is because during the cycling process. Under the action of the upper limit constant pressure load, the mudstone is compacted by extrusion. Therefore, the residual strain before the interval is larger than the residual strain after the interval.

Meanwhile, it can be found that in the triaxial intermittent fatigue test, a negative deformation phenomenon occurs after the interval. This is because a large amount of elastic energy is stored inside the mudstone during loading and constant pressure loading. In the unloading stage, the release of elastic energy led to the recovery of deformation and the generation of negative deformation.

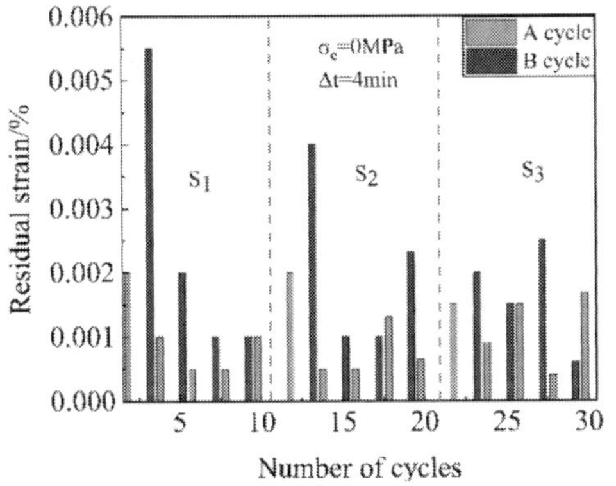

(a)

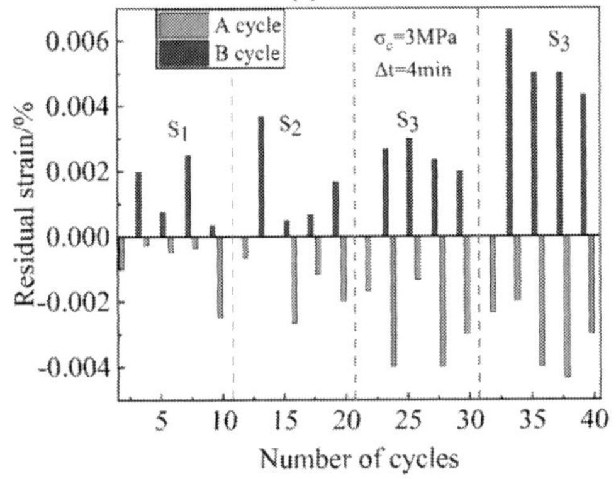

(b)

Fig.6 GUIF test residual strain (a)σ_c=0MPa,Δt=4min, (b)σ_c=3MPa,Δt=4min

To reflect the development law of residual strain under different stress levels more clearly, graded uniaxial intermittent fatigue test is taken as an example. The average residual strain of mudstone under each stress level was counted. As shown in Fig. 7.

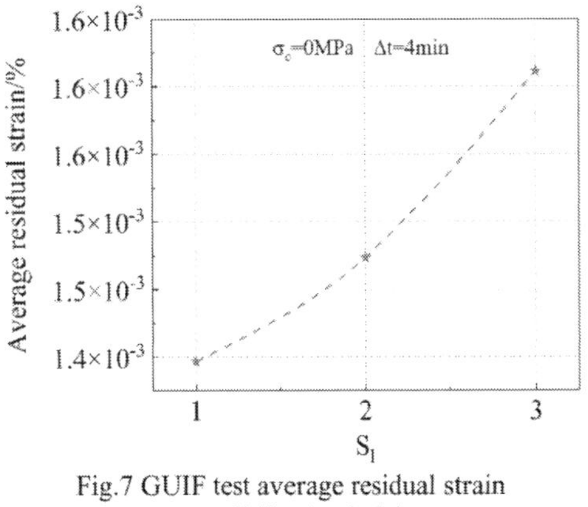

Fig.7 GUIF test average residual strain (σc=0MPa,Δt=4min)

As can be seen in Fig. 7, the average residual strain at each stress level tends to increase nearly as an exponential function as the stress level increases. This is due to the fact that the stress amplitude increases as the stress level rises, and the loading action increases the destructive capacity of the mudstone, thus generating more irreversible deformation, which is also known as residual strain. So the average residual strain increases with increasing stress level.

4. Conclusion

During the operation of underground gas storage reservoirs, the surrounding rock experiences discontinuous cyclic loading action. To investigate the effect of intermittent action on the fatigue mechanical properties of mudstone, this paper sets up continuous fatigue test and graded uniaxial intermittent fatigue test as a control. Considering the horizontal stress on the rock under the actual burial depth, this paper sets graded triaxial intermittent fatigue test and graded uniaxial intermittent fatigue test as a control. By analyzing the data obtained from the tests, it is found that the fatigue mechanical properties of mudstone under the intermittent action and confining pressure

It has certain reference value for the construction and stable development of gas storage. For example, when setting the frequency of gas injection and extraction, the constant load period of the upper stress limit should be extended appropriately. And the gas confining pressure injected in the cavity should be controlled in a reasonable range to prevent the collapse of the surrounding rock.

For this indoor test, there are mainly the following conclusions:

1. the fatigue life of mudstone is enhanced by the intermittent effect during the fatigue process. The presence of confining pressure under the same experimental conditions also benefits the fatigue life of mudstone.
2. the average modulus of elasticity of mudstone decreases with increasing stress level during cyclic loading, and both intermittency and confining pressure help to enhance the modulus of elasticity of mudstone.
3. in graded intermittent fatigue tests. In both uniaxial and triaxial tests, the residual strains generated by the post-

interval cycles are smaller than those generated before the intervals.

4. In the 3MPa graded interval fatigue test, negative deformation occurs in the post-interval cycle due to the release of elastic energy.

References

1. ZHU Jiang. 《Rocks and Landforms. 》 Chongqing: Chongqing University Press,2014.11
2. LIU Te-hong,LIN Tian-Jian. Engineering Design Theory and Construction Practice of Soft Rock [M]. Beijing: China Architecture and Building Press,2001
3. WANG Ke, PAN Dan-Guangli Wen-Yan, CHEN Fan. Experimental study on damping characteristics of mudstone and sandstone under graded cyclic loading[J].Journal of Central South University (Science and Technology), ,2019,50(12):3046-3054.
4. CHEN Fang-ming, HU Quan-guang, NING Guang-Zhong. Silt mudstone stress relaxation characteristics under tri-axial stress condition[J]journal of Shandong university,2017,47(03):125-129.
5. ZHANG Xiang-Dong, YIN Xiao-Wen, FU Qiang.22/5000 Study on triaxial creep characteristics of purple mudstone under graded loading [J].JOURNAL OF EXPERIMENTAL MECHANICS, 2011, 26(01):61-66.
6. HE Ming-ming, LI Ning, CHEN Yun-sheng, ZHU Cai-hui. An experimental study of dynamic behaviors of rock under stepwise cyclic loading[J]. Rock and Soil Mechanics,2015,36(10):2907-2913. DOI: 10. 1285/j.rsm.2015.10.021.
7. ZHANG Qiang, WANG Jun-bao, SONG Zhan-ping, FENG Shi-jin, ZHANG Yu-wei, ZENG Tao. Microstructure variation and empirical fatigue model of salt rock under cyclic loading[J].Rock and Soil Mechanics,2022,43(04):995-1008. DOI: 10. 16285/ j.rsm.2021.1356.
8. CHEN Wen-wu, YUAN Peng-bo, LIU Xiao-wei. Experimental study on creep characteristics of red bed soft rock under graded loading[J].Chinese Journal of Rock Mechanics and Engineering, 2009,28(S1):3076-3081.
9. LU Gao-Ming, LI Yuan-hui1,ZHANG Xi-Wei, LIU Jian-Po. Fatigue deformation characteristics of yellow sandstone under cyclic loading[J].Chinese Journal of Geotechnical Engineering, 2015, 37(10):1886-1892.
10. XU Yin, LI Cheng-Jie, ZHENG Qiang-Qiang, NI Xian, WANG-Qian-Qian. Analysis of energy evolution and damage characteristics of mudstone under cyclic loading and unloading[J]. Chinese Journal of Rock Mechanics and Engineering, 2019,38(10):2084-2091.DOI:10.13722/j.cnki.jrme.2019.0153.
11. CAI Guo-Jun, CHEN Yu-hang, ZHONG Chuang, ZHOU Yang. Study on Dynamic Characteristics of mudstone under cyclic loading and unloading [J]. Gansu Water Resources And Hydropower Technology, 2019, 55(05): 20-24. DOI: 10.19645/j.issn2095-0144.2019.05.006.
12. GUO Yin-tong, ZHAO Ke-lie, SUN Guan-hua, YANG Chun-he, MA Hong-Ling, ZHANG Gui-mi. Experimental study of fatigue deformation and damage characteristics of salt rock under cyclic loading[J]. Rock and Soil Mechanics, 2011, 32(05):1353-1359.DOI:10.16285/j.rsm.2011.05.009.
13. LI Zong-ze, JIANG De-yi, FAN Jin-yang, CHEN Jie, LIU Wei, WU Fei, DU Chao, KANG Yan-Fei. Experimental study of triaxial interval fatigue of salt rock[J].Rock and Soil Mechanics,2020,41(04):1305-1312+1322.DOI:10.16285/j.rsm.2019.0711.
14. JIANG De-Yi, FAN Jin-yang, CHEN Jie, YANG Chun-he, CUI Yao. Influence of interval fatigue tests on fatigue characteristics of salt rock[J].Chinese Journal of Geotechnical Engineering, 2016, 38(07):1181-1186.
15. FANJY,CHENJ,JIANGDY,etal.Fatiguepropertiesofrocksaltsubjectedtointervalcyclicpressure[J].InternationalJournalofFatigue,2016,90(9):109-115.
16. TAN Yun-zhi1,HU MO-zhen, ZHOU WEI-tao, ZUO Qing-jun ,WANG Hong-xing, YU Bo. Effects of drying-wetting cycle and loading on compressive property of mudstone[J].Rock and Soil Mechanics, 2016, 37(08): 2165-2171. DOI: 10.16285/j.rsm.2016.08.006.
17. REN Song, BAI YUE-ming, JIANG De-yi, YANG Chun-he,Chen Jie. Experimental Study on Influence of Temperature on Fatigue Characteristics of Salt Rock [J].Chinese Journalof Rock Mechanics and Engineering,2012,31(09):1839-1845.
18. WANG Rui-Hong, LI Jian-Lin, JIANG Yu-zhou, CHEN Xing. Experimental research on influence of cyclic loading and unloading on rock mass residual strength[J]. Chinese Journal of Rock Mechanics and Engineering,2010,29(10):2103-2109.
19. YANG Yong-jie, SONG Yang, CHU Jun. Experimental study on characteristics of strength and deformation of coal under cyclic loading[J].Chinese Journal of Rock Mechanics and Engineering, 2007(01):201-205.
20. ZHU Zhuo-hui, FENG Tao, GONG Feng-qiang, YE Zhou-yuan, YU Zhi. Experimental study on mechanical properties of coal and rock mass with graded cyclic loading and unloading [J]. Journal of Central South University (Science and Technology),2016,47(07):2469-2475.
21. HE Ming-ming, Li Ning, CHEN Yun-sheng, ZHU Cai-hui. Experimental study on damping behavior of soft rock under cyclic loading [J]. Chinese Journal of Solid Mechanics, 2016, 37(01):74-82.DOI: 10.19636/j.cnki.cjsm42-1250/o3.2016.01.005.
22. FENG Chun-lin, WU Xian-qiang, DING De-xin, WU Yong-feng. Study on fatigue characteristics of white sandstone under periodic loads [J]. Chinese Journal of Rock Mechanics and Engineering, 2009,28(S1):2749-2754.

23. ZHANG Li-ming, YANG Ge-xiao, ZHANG Tao, LUO Jun-hui, HUANG Hai-feng. Experimental study on evolution law of mechanical properties of carbonaceous mudstone under dry-wet cycle [J]. Highway,2021,66(10):340-345.
24. ZENG Ling, LUO Jin-tao, HOU Peng, BIAN Han-bing. Dry-wet circulation under the action of pre disintegrating carbonaceous mudstone fracture rule and strength characteristics [J]. China journal of highway and transport, 2020 (9) : 1-11. DOI: 10.19721/ j.carol carroll nki. 1001-7372.2020.09.001.
25. LU Xiang, ZHOU Wei, CAI Qing-xiang, LI Ming, LUAN Bo-yu, LIU Fu-ming. Shale mechanical properties under the freeze-thaw cycle and mesoscopic fracture mechanism study [J]. Journal of mining and safety engineering, 2021, 38 (4) : 819-829. The DOI: 10.13545 / j.carol carroll nki jmse. 2020.0633.
26. SHEN Pei-wu, TANG Hui-ming, WANG Ding-jian, HE Cheng, ZHANG Ya-hui. Badong group purple mudstone dry-wet circulation collapse characteristics experimental study [J]. Rock and soil mechanics, 2017, 38 (7) : 1990-1998. The DOI: 10.16285 / sm j.r. 2017.07.019.
27. YANG Jian-lin, WANG Lai-gui, LI Xi-lin, ZHAO Guo-chao. With water - air drying cycle under the action of mudstone fracture micro mechanism research [J]. Journal of rock mechanics and engineering, 2014 (S2) : 3606-3612. The DOI: 10.13722 / j.carol carroll nki jrme. 2014. S2.027.
28. KONG Xiang-hui, JIANG Guan-Lu, ZOU Zu-yin, LI An-Hong. Study on dynamic and pavement performance of red mudstone under cyclic load [J]. Chinese Journal of Rock Mechanics and Engineering,2013,32(09):1813-1819.
29. WANG Zhi-meng, JIANG Guan-lu, WEI Yong-xing, HU An-hua. Experimental study on cyclic loading of red mudstone Subgrade of Reach line [J]. Chinese Journal of Geotechnical Engineering, 2008, 30(12):1888-1893.
30. WANG J G, SUN Q L, LIANG B, et al. Mudstone creep experiment and nonlinear damage model study under cyclic disturbance load[J]. Scientific Reports, 2020, 10(1): 9305

Discussion on ecological restoration strategy of water conservancy project -- A case study of Shirase Reservoir in Quanzhou City

Shizhe Fan *, Mingge Li, Jingwen Huang, Yuqing Cheng, Wenye Zhou

School of advanced manufacturing, fuzhou university, Fuzhou, China

Abstract: Because of the water conservancy project construction to the environment has brought a lot of damage, the state for environmental pollution and damage problems began to carry out a comprehensive restoration, for the relevant environmental protection and restoration policies are also improving, with the implementation of relevant measures, ecological water conservancy projects will play a vital role in environmental governance. In order to put forward suggestions for ecological restoration of white rose reservoir, this paper puts forward suggestions for ecological restoration from the geographical location of the water conservancy project, the measures in the construction process and the operational measures after completion, so as to restore the ecological role played by the ecological area before the completion of the water conservancy project and reduce the secondary pollution and damage to the environment in the operation process after it. This paper discusses the importance of ecological water conservancy project in environmental governance, and takes Shirase reservoir as an example, puts forward the ecological restoration suggestions for the water conservancy project. In order to restore the ecological function of the area before the water conservancy project and reduce the secondary pollution and damage in the operation process, it is suggested to improve the area from three aspects: geographical location, construction process and operation measures after completion. Through the implementation of ecological restoration measures, it will help to realize the harmonious symbiosis between man and nature.

Key words: Ecological water conservancy project; Pollution control; Ecological restoration; Composite engineering system; Shirase reservoir.

1. Introduction

Since the founding of the People's Republic of China, the emphasis has been placed on the development of economic level for most of the time, resulting in the great destruction of ecological environment by engineering construction, and these damages have also begun to cause adverse effects on human life. At present, the self-purification function of the environment has been unable to meet the needs of the environment. During the construction period of Bailai reservoir, due to some unreasonable construction measures, coupled with the lack of timely restoration, there have been serious ecological problems in the water conservancy project area. The development of the composite engineering system of water conservancy engineering and ecological engineering will better solve the ecological environment problems. As one of the important means of environmental governance, ecological water conservancy engineering will play a vital role in the future development. This paper takes the white ase reservoir as an example, discusses the impact of water conservancy projects on the ecological environment and its causes, and puts forward targeted ecological restoration suggestions, in order to provide reference for the construction of related water conservancy projects and ecological environmental protection, and puts forward suggestions on the ecological restoration of the current reservoir with specific cases and implementation measures.

2. Project overview and the background of the formation of ecological environment problems

2.1 Geographical location of the project

The dam is located about 600 meters downstream of the confluence of Xixi and Baiheng Creek in the Jinjiang River Basin. The area where the project is located has plenty of sunshine and abundant precipitation. The terrain around the project is medium and low mountain, with deep mountain gullies, gentle upper part of the

* Corresponding author: 1220095345@qq.com

valley slope and steep lower part. Human disturbance activities are frequent in the area around the dam, and there is almost no native zonal vegetation distribution, and the existing vegetation species are single, and most of the vegetation varieties are artificial coniferous forests. The ecological diversity of the ecosystem in the project area is relatively simple, and the resistance of the ecosystem is weak. On the other hand, the basic engineering and hydropower facilities in Xixi Basin have had a great impact on the river ecosystem in the basin.

2.2 Main contents of the project

The development objective of the project is mainly water supply and drainage, comprehensive flood control, hydro power generation, and creating conditions for improving the level of agricultural and ecological water use in the downstream basin. Focus on the construction of diversion system, drainage bottom hole, RCC gravity dam, generator room and plant area construction, fish equipment system engineering, boost voltage substation project, reservoir waterproof leakage construction and flood control channel between the left and right DAMS and other infrastructure projects.

2.3 Main causes of ecological pollution caused by the project

The pollution caused during the construction of Shirase reservoir is mainly non-point source pollution, that is, non-point source pollution, which can be defined as the water pollution caused by dissolved or solid pollutants flowing into the receiving water body from non-specific places through the runoff process. Different from point source pollution, there is no fixed pollution discharge point. Non-point sourcepollution mainly includes urban non-point source pollution and agricultural non-point source pollution. Typical urban non-point source pollution refers to stormwater runoff pollution, which is specifically manifested as river discharge pollution of municipal pumping station. Agricultural non-point source pollution includes farmland runoff and leakage, rural surface runoff, rural domestic sewage, rural solid waste, livestock, poultry and aquaculture and other pollution.

3. Ecological impact and causes of the project

3.1 Ecological impact during the construction period

Through the ecological investigation of the local waters, in the area from Shirase Hydropower Station dam to Jianan hydropower plant dam, due to the implementation of multi-stage water conservancy and hydropower development, there is no water area where fish can spawn and reproduce, and only some hydropower station tail river area can meet the spawning requirements of fish. There is no effective management of domestic sewage in the urban area of the intersecting water basin in the project area, and this part of sewage, rural sewage and a small amount of industrial production sewage directly enter the drainage area of the project construction area, causing the water environment in the project area to be polluted to a certain extent. At the same time, the natural environment in the project area is also greatly affected by human's development and construction of the environment and human's daily life. Non-scientific tree cutting and vegetation destruction, tea garden reclamation and cultivated land reclamation, farmers' cultivation and livestock breeding activities have increased the soil and water loss in the project construction area.

3.2 Ecological impact during the project operation period

The huge noise and dust pollution generated by the project will affect the ecology of the river channel, and the rare protected animals need to be driven away before the reservoir is flooded, and these animals may have difficulty living or even die because they do not adapt to the new environment after being driven away. If the driving method is not reasonable, it may also directly cause the death of animals. After the operation of the project, the river flow will be greatly changed, especially the flow of the downstream river, which is a big survival risk for the organisms that depend on the river to survive. The number of aquatic animals and plants that do not adapt to the change of water flow will decrease greatly, and even become extinct. A reservoir is a closed ecological environment. The food chain and food web of this ecosystem are simple, and the material exchange and energy transfer within the system are not complicated. Therefore, its resistance stability and resilience stability are relatively low, and it is easy to be disturbed by external factors, especially human factors, which is not conducive to the growth of biodiversity, and will make the local water ecosystem's ability to resist the invasion of foreign organisms decline. The affected area of reservoir inundating involves fifteen villages, two township government locations and one forest farm in five towns and two counties of Quanzhou City. The cultivated land in the inundated area can no longer be developed.

4. Environmental impact and causes of the project

4.1 Environmental impact during construction period of the project

The RCC gravity dam is used in all the barrage DAMS of this project, and the excavation work in the dam site area of Haase reservoir is very heavy. Part of the land excavated in the barrage area is used as cofferdam by the construction enterprises, and the efficiency and benefit of the backfilling work of the earthwork and rock works are low, resulting in a lot of project waste slag and waste soil in the hub project area. At the same time, due to the special geological conditions of the river in which the project is located, the slag dump in the project area is located below the dead water level line of the reservoir inundate area, thus forming a reservoir type slag dump. The slag dump is generally an important area of soil and

 https://doi.org/10.1051/e3sconf/202447801022

water encroachment, which has the characteristics of great harm, difficult treatment, large loss and strong sudden occurrence. The slag dump is deposited on the shore, and when the prevention measures are not appropriate during the construction period, the slag dump will collapse or the slag dump will be washed away by the flood, which will lead to strong geological disasters such as mountain flood and debris flow. It may also block rivers in nearby waters, affect flood discharge, and even endanger project construction. As mentioned above, the project area is already a serious area of soil and water loss, and the construction period will accelerate soil and water loss. The dust and noise generated during the construction period will also affect the normal life of the surrounding residents. There is a stream with a narrow channel in the upper reaches of Shirase Reservoir. The area along the stream is rich in coal resources, and a considerable amount of coal gangue is discharged into the river every year. The upstream river is narrow and the water flow is small, so during the construction of the project, if the upstream slag is not treated well, coupled with the project itself, it is easy to cause debris flow disasters.

4.2 Environmental impact during the project operation period

After the construction of Shirase Reservoir, the upstream and downstream water flow speed will be changed, and the discharge volume will be reduced during the storage period. Therefore, the construction of the reservoir dam will change the upstream and downstream hydrological regime of the dam, and then bring harm to the ecological environment of the upstream and downstream basins. The construction of the reservoir will also change the water temperature, which will bring unfavorable factors to the local crop planting and the growth and reproduction of aquatic organisms, and the harm to the water body will also cause a substantial increase in harmful substances in the water body, thus causing ecological harm to the environment. At the same time, after the operation of Shirase Reservoir, a large amount of water will be stored, which will lead to the increase of rainfall in the area where the reservoir is located and in the surrounding areas, and change the original rainfall variation law in the basin, which will have a certain negative impact on the local crop yield. To a certain extent, the construction of the project will reduce the impact of geological disasters such as debris flow and flash floods on agricultural production and local environment, but it will also make the land near the reservoir salinized and even swamped. This will also lead to a decrease in the fertility of the land in the area, and if the situation is serious, it will even lead to this part of the land is no longer suitable for cultivating food crops and cash crops, and farmers have to waste this land. After the construction of water blocking structures, water storage reservoirs will be established in the upper reaches of the project. Due to the rise of the water level, the water flow rate of Shirase Reservoir will gradually decrease and the waterretention time will become longer. In other words, the flow rate in the reservoir decreases, which reduces the exchange rate between the water body and the air interface and the diffusion capacity of pollution, and the reoxygenation capacity decreases. The decre ase of water flow in the reservoir will lead to the improvement of water permeability, which is conducive to the photosynthesis of algae, and then lead to the large-scale growth and reproduction of algae, which will cause the excessive content of N, P, K and other elements in the water body of the reservoir, resulting in the eutrophicationof the water body and the formation of bloom, which will affect the ecological landscape and ecological environment. If not controlled, it will even lead to the collapse of the ecosystem of the reservoir. In addition, since the storage system not only stores the excess water in flood season, but also intercepts the base flow of the river regardless of flood season, it will cause the decrease of the upstream and downstream horizontal water volume, which will lead to the decrease of the horizontal water level line of the upstream and downstreamrivers and the horizontal water level line of the lake surface, the wetland area in the water area will decrease, and the carrying capacity of the upstream and downstream wetland ecosystem will decline. Once the leakage problem occurs when the reservoir is impounding, it will lead to great changes in the downstream hydrology. When the reservoir is filled, the pressure of the water increases, which causes the crustal stress to rise. At the same time, water will also enter the fault, making the fracture edge smoother and less friction, which will increase the pressure of water flow in the pores of the formation, and the reservoir bank may collapse. Therefore, after the completion of Shirase hydropower Station project, it may cause special geological natural disasters such as earthquake and landslide on the mountain. After storage, the water level of Shirase reservoir will rise, which will lead to the decline of the shear resistance of the soil on the surrounding bank slope, which may cause rock collapse, rock collapse, and even damage of some rock mass in the surrounding mountain. In order to ensure the resistance and stability of the dam, the project will artificially inject cement concrete and hypothetical steel bars into the mountains around the dam, which can stabilize the originally unstable mountains and reduce the possibility of geological disasters in the mountains in the area where the dam is located. Engineering construction enterprises can design the project during the operation period to create an ecological landscape environment and hydrophilic engineering (such as biological bank protection, ecological natural viewing platform, etc.) in harmony with the surrounding nature, so as to improve the landscape value of the environment and the vitality of the ecosystem.

5. Comprehensive evaluation of ecological environment and corresponding solutions

5.1 Comprehensive ecological environment evaluation during the construction period of the project

Since many small and medium-sized hydropower stations have been built on the tributaries of Xixi Trunk in Jinjiang River Basin, which has an impact on fish migration, the project has gradually changed the hydrological condition of the downstream river, and then squeezed the living space of the downstream fish. The living space of local aquatic animals and plants is a kind of occupation, which will affect or even threaten the income of aquatic products and the biodiversity of local aquatic animals and plants. During the construction period, a certain range of vegetation will be destroyed, a certain range of geological structure will be affected, and the amount of mud and sand in the river will be increased. The environmental damage will be unspeakable.

Considering that the project has changed many influencing factors such as local rainfall, soil salinity and river temperature, the project will certainly have an impact on the local natural environment, which is mainly reflected in the loss of agricultural products and fisheries, the reduction of local biodiversity, and the destruction of the original natural ecological landscape.

5.2 Environmental impact assessment during the operation of the project

After the completion of the project, a certain amount of dust and noise will be generated, which will definitely change the original ecological situation and affect the lives of surrounding residents. The project will cause the difference of climate and environment to expand gradually, and the balance of ecological environment will be destroyed. As mentioned above, the change of river hydrological situation determines that the project must pay attention to the change of ecological environment. If we do not pay attention to the ecological protection during the operation period, the consequences brought by the project must be irreparable. During the operation period, the biggest change of the project on the ecosystem is actually to reduce the stability of the original ecosystem and biodiversity, and the loss depends on how the project deals with the relationship between the construction and nature.

After the completion of the project, the water shortage situation in the downstream area will be completely changed, and the local agricultural and industrial development will be greatly promoted. After the project starts operation, it can greatly improve the flood control capacity of the downstream area, which can not only protect the downstream ecological environment, but also protect people's life and property safety.

5.3 Mitigation of negative impacts brought by the project and corresponding measures for ecological restoration

5.3.1 *Construction period of the project:*

(1)Improve the implementation plan of pollution remediation of construction ente rprises, and strictly implement the management measures. Construction enterprises must establish a systematic monitoring system to evaluate the severity of water pollution in river basins, analyze quantitatively and qualitatively the cumulative factors and the risk of environmental pollution to water bodies in construction areas, and implement pollution control measures "tailored to local conditions".

(2) Accelerate the establishment of an ecological guarantee mechanism to ensure that the construction of pollution control projects is completed on schedule and in quantity. Organize the implementation of the comprehensive remediation plan for the upstream and downstream pollution sources of the project, and gradually determine the specific plan for all the comprehensive remediation work and the person in charge of the implementation work, ensure the regularquantitative testing, and coordinate the relevantpersonnel to carry out the testing. Before the reservoir is impounding, complete various tasks strictly according to the comprehensive renovation implementation plan that has been prepared fo r construction, and organize special acceptance inspection to ensure that the project does not cause major harm to local water bodies.

(3) Strictly implement the major principles of water resources protection. The construction of Shirase Water Conservancy project has increased the use of urban water resources, so the quantity of water diversion must be strictly controlled. At the same time, the saved water resources should be prioritized to improve the ecological water use of the river, and must not preempt ecological water use.

(4) Strictly implement ecological protection policies and measures. Measures such as artificial breeding and release of fish species and construction of fish ecological habitats should be taken to protect fish in the basin. Before water flow interception, the local management station should build a fish breeding and releasing station, establish a tracking and monitoring system, and carry out dynamic evaluation, so as to rationally control the species of fish breeding and releasing and the release specifications according to the evaluation content.

(5) Strictly control the discharge of engineering waste to reduce the pollution of dust to the environment during construction. Dust reduction measures should be adopted during the transportation, temporary storage and handling of bulk materials such as earthmoving, building materials and lime. Storage yard, mixing station is best set away from residential areas, construction equipment should be sealed and installed dust removal device. Strengthen the management of construction vehicles to avoid the generation of vehicle

noise, dust pollution caused by falling objects and other problems. The municipal solid waste formed during construction must be sorted and disposed of, but it is best to make secondary utilization, and the waste that cannot be utilized should be disposed of according to relevant laws and regulations. It is strictly prohibited to arbitrarily discharge construction waste into water bodies.
(6) Implement measures for water and soil conservation and protection in the construction area. Strengthen the forest management and protection near the project area, and gradually make the forest species structure, age group composition and tree structure of the forest farm in the surrounding area more scientific and reasonable through forest land operation, so as to achieve the persistence and development of ecological environmental efficiency. Avoid the loss of local forest farm resources due to the occupation of forest farm for construction.

5.3.2 Project operation period:

(1) The drainage protection measures of ecological construction flow must be strictly implemented. According to the ecological planning during the construction and operation period, the flow rate of ecological construction should meet the environmental requirements of the river under the dam. During operation, ecological drainage bypass pipe and turbine unit can be used to control ecological drainage velocity, so as to realize immediate monitoring and dynamic monitoring of ecological velocity.
(2) It is recommended to use stratified water diversion of reservoir. The water body of the power station will cool down rapidly after discharge. It is suggested to carry out special research on classified water diversion, and the classified water diversion equipment must be built at the same time as the main project. During operation, the downstream water temperature should be monitored by a comprehensive tracking system.
(3) Fully implement environmental protection measures. Formulate environmental cleaning plans for reservoirs, focusing on the disposal of sediment in reservoir areas and tributaries that exceed the standard of pollutants. To do a good job in the restoration and improvement of the regional environment near the project, and to use the obstruction diversion project to strictly meet the standards of the industrial wastewater generated by the Pantian Iron Mine and then to the Jiulong River basin for local crop irrigation. To carry out the construction of Shuangxi constructed wetland downstream of Pantian Iron Mine, and carry out comprehensive treatment of sediment from Pantian Iron Mine to the dam site watershed.
(4)Follow up and maintain the animal and plant protection areas built during the construction period. If the animal and plant reserves lack personnel maintenance, the previous ecological protection will be wasted, only continuous follow-up protection can achieve the purpose of species ecological protection. Plants that play important ecological functions will be replanted to ensure the normal operation of the original ecosystem. Clean up the construction waste generated during the construction period, and send relevant personnel to carry out regular and quantitative ecological inspection of the original area after the completion of the project, and remedy the ecological problems in time if found.
(5)Select indicators to establish a comprehensive evaluation index system of water conservancy project operation benefits. [15] The composite weight method is used to calculate the weights of different indicators, and the evaluation model based on the cooperation degree between different benefits is established to comprehensively evaluate the operating benefits of Shirase reservoir. The established evaluation model can comprehensively consider the correlation degree between the three sub-benefits, which solves the limitation that the traditional evaluation model can not consider the correlation degree between the benefits, and the comprehensive evaluation results are more reasonable.

6. Conclusion

This paper mainly discusses the environmental damage caused by the construction of water conservancy projects, and the measures taken by the state to repair the environmental pollution and damage. In view of the ecological restoration of Shirase Reservoir, suggestions are put forward from the aspects of geographical location, construction process and operation measures after completion, so as to restore the ecological function of the reservoir before its completion and reduce secondary pollution and damage in the operation process. Through the implementation of ecological water conservancy projects and restoration measures, it will help to solve the ecological environment problems and realize the harmonious coexistence of man and nature. To sum up, the ecological restoration project of Shirase Reservoir is of great environmental significance and social value. During the construction period, due to unreasonable construction measures and lack of timely restoration, more serious ecological problems have appeared in the water conservancy project area. Therefore, suggestions for ecological restoration are put forward to restore the ecological role played by the ecological area before the completion of the water conservancy project and reduce the secondary pollution and damage to the environment in the operation process. In the ecological restoration project of Shirase reservoir, a variety of measures should be taken, including restoring vegetation, improving water quality, and enhancing the resistance and stability of the ecosystem. According to different pollution sources and pollutants, corresponding treatment measures should be taken to realize the comprehensive restoration of the ecological environment. In the future development, attention should be paid to the organic combination of water conservancy engineering and ecological engineering to achieve the harmonious coexistence of human activities and the natural environment. At the same time, the supervision and management of water conservancy projects should be strengthened to avoid secondary pollution and damage to the ecological environment. Finally, we should realize that ecological environment protection is a long-term and arduous task, which requires the joint efforts of the whole society. Only by strengthening policies and regulations, scientific and technological innovation, public participation and other aspects of comprehensive management, can we achieve the continuous improvement of the ecological environment and the sustainable development of human society.

References

1. Anonymous. Progress of Baiase Water Conservancy Project in Quanzhou, Fujian [J]. Water Pump Technology, 2016.
2. GONG H J. Ecological environment management and restoration plan of Bai Se Reservoir engineering area [J]. 2022(6). (in Chinese)
3. Fujian Development and Reform Commission. National Development and Reform Commission approved the feasibility study report of Quanzhou Bai Se Water Conservancy Project [R]. Quanzhou City, Fujian Province: Fujian Development and Reform Commission, 2018.
4. WANG Congmin. Ecological flow guarantee measures for Quanzhou Bai Se Water Conservancy Project [J]. Water Conservancy Science and Technology, 2021(1):3.
5. Lin Rong. "double insurance" tank drought-resistant multiply Rio - quanzhou white wrasse water conservancy hub project starts sidelights [EB/OL]. [] on January 2, 2020 in 2020, http://www.hssczl.net/2020-01/02/content_6049884.htm.
6. LU Jing. Study on environmental protection measures for resettlement of migrants in Baiase Water Conservancy Project in Quanzhou, Fujian Province [J]. Northern Environment, 2019.
7. Anonymous. Notice of Fujian Provincial People's Government on the prohibition of new construction projects and resettlement population within the land requisition of Quanzhou Bai Se Water Conservancy Project [J]. Bulletin of Fujian Provincial People's Government, 2015.
8. Anonymous. Reply to the Environmental Impact Report of Quanzhou Bai Se Water Conservancy Project in Fujian Province [J]. Official Document of the Government, 2017
9. Anonymous. Letter of Quanzhou Environmental Protection Bureau on the approval of the Environmental Impact Report of Quanzhou Bai Se Water Conservancy Project Grade Road relocation and Reconstruction Project [J]. Government Official Document, 2016.
10. ZHANG M. Flood impact analysis and countermeasures of reservoir type slag dump of water conservancy project [J]. Subtropical Soil and Water Conservation, 2018, 30(2):5. (in Chinese)
11. Huang Runxiong. Discussion on the potential harm of coal gangue disaster in Qidouxi to Bai Se Reservoir and its prevention and control measures [C]// Straits Science and Technology Expert Forum and Cross-Strait Geological Disaster Prevention Symposium. Geological Society of China; Geological Society of Fujian Province, 2012.
12. qiu-sheng Lin, cathleen. Water conservancy project compensation measures to analyze the impact on the environment and its engineering [J]. Water power technology, 2007, 33 (1) : 8.
13. WANG Z Y. Ecological environment impact and control measures of water conservancy projects [J]. Journal of Yellow River Conservancy Technical Institute, 2011, 23(004):7-9. (in Chinese)
14. Sheng Lianxi, Hou Wenli. [14] Effect of cooling system in power plant on coiling load of shuttle fish and prawn larvae [J]. Chinese Journal of Environmental Sciences,1994,14(1): 47-55.
15. Wang Guo-kui. Research on comprehensive evaluation model of water conservancy project operation benefit based on cooperation degree [J]. Water Conservancy Planning and Design, 2022(010):000.]
16. Ge Mingkun. Review of non-point source pollution treatment theories and measures in China [J]. Water Conservancy Planning and Design, 2020(3):5.]

Geological Storage Conditions and Potential Analysis of Carbon Dioxide in the High-steep Fold Belt in Eastern Sichuan Basin

Jinhua Luan [1,2], Ruigang Zhang [1], Haitong Zhao [1,*], Xiong Zhang [3], Lingling Mao [1], Yi Zhang [1], Feng Hu [1], Heming Tian [1], Yi Dong[1], Liangmei Jiang[1], Fei Chen[1]

[1] Chongqing Key Laboratory of Major Geological Events and Effects of Resources and the Environment (Chongqing Institute of Geology and Mineral Resources), 401120, China
[2] School of Resources and Geosciences, China University of Mining and Technology, China
[3] Key Laboratory of Shale Gas Exploration, Ministry of Natural Resources (Chongqing Institute of Geology and Mineral Resources), 401120, China

Abstract. Carbon sequestration refers to the technology of capturing carbon and storing it safely instead of directly discharging CO_2 into the atmosphere. Rock salt is a promising CO_2 reservoir, and the underground rock salt structure formed by salt cavern reservoirs in geological history has a natural sealing function. Storing carbon dioxide in salt caves can ensure that carbon dioxide is stably stored underground for a long time. However, at present, there are few studies on carbon sequestration in abandoned salt caves. This study analyses the characteristics of carbonate-evaporite strata in the high fold zone of eastern Sichuan and qualitatively evaluates Triassic carbonate-evaporite layers formed in the depth range of 800-3000 m in various sedimentary basins in Chongqing. A large set of carbonates-evaporites are developed in Chongqing, and the carbonate rocks near evaporites have low porosity and permeability, which can be used as CO_2 storage caprock in salt cavities. Thick evaporated strata can also provide a good cover for CO2 storage in abandoned salt caves, which has great potential space resources.

Key words: Salt cavern reservoirs; carbonate reservoirs; Triassic system; Eastern Sichuan basin

1. Introduction

Carbon dioxide geological storage technology (CCS) is regarded as a direct and effective means to reduce emissions. This technology captures CO_2 originally discharged into the atmosphere and injects it into closed underground reservoirs after treatment, thus reducing the emission of CO_2 into the atmosphere and slowing down air pollution and the greenhouse effect. The ultimate goal of CCS is to isolate CO_2 from the atmosphere for a long time and safely to reduce the concentration of greenhouse gases in the atmosphere. Therefore, CO_2 storage is the most critical link of CCS, and the analysis of storage potential and storage cost of storage sites has become an important part of CCS evaluation. CCS aims at large-scale CO_2 emission reduction. For the early implementation of CCS, from the perspective of saving transportation costs, the storage potential of a storage site should be at least greater than the total emission of a large coal-fired power plant in 20-30 years (40×10^6 t-60×10^6 t); otherwise, CCS is of little significance to greenhouse gas emission reduction. Second, the reservoir is required to be well sealed, that is, the caprock overlying the reservoir is impermeable or extremely low permeability rock, to prevent the injected CO_2 from leaking through the caprock, causing many CCS projects to fail. At present, CCS technology has been widely studied by researchers around the world. Safe storage of CO_2 is the ultimate goal of CCS, and the feasibility of geological storage determines the feasibility of CCS technology.

Under the situation that China's coal-based energy consumption structure is difficult to change in the short term, developing geological storage of carbon dioxide is an effective measure to realize China's carbon emission reduction commitment. In recent years, the geological storage of carbon dioxide has developed rapidly in the fields of regional investigation and evaluation, key technology research and engineering demonstration. The geological utilization and storage technology of carbon dioxide has gradually become a system, but it is still in the demonstration research stage. It is of great significance for the planning and implementation of carbon capture, utilization and storage in China to continue the investigation and evaluation of carbon dioxide geological storage and innovate the technical methods. It has been less than 10 years since China achieved the peak carbon dioxide emissions goal, and only 30 years since peak carbon dioxide emissions achieved the goal of carbon neutrality. Compared with developed countries, it is shorter, more extensive, more difficult and extremely

* Corresponding author: zhaohtxyl@163.com

arduous to achieve the long-term goal of peak carbon dioxide emissions and carbon neutrality. Chongqing's emission sources are mainly concentrated in the cement and thermal power industries. Based on the actual situation in Chongqing, the total amount of coal decreased during the 14th Five-Year Plan period, and the process of low-carbon transformation gradually accelerated. Using the evaluation index system and evaluation technology method of carbon dioxide geological storage potential and suitability established in China, the potential and suitability of carbon dioxide geological storage in the eastern Sichuan Basin were evaluated. This study focuses on eastern Sichuan basin and discusses the carbon sequestration potential of Triassic carbonates-evaporite strata in the Dianjiang salt basin.

2. Geological settings

The eastern part of the Sichuan Basin is a large-scale continental clastic sedimentary basin that contains oil, gas, salt and brine. The reservoir types of oil and gas are mainly fluvial delta facies and shore-shallow lake facies sand bodies. Rock types are mainly carbonate rock, shales and evaporites. Generally, these evaporated rocks have good physical properties and good traps, which is a potential reservoir basin in which the Jialingjiang Formation has the potential for sealing and is also a continental sedimentary basin with high salinity. One basin in Sichuan and Chongqing is larger than 100,000 km^2, one in 100,000-10,000 km^2, two in 100,000-100,000 $km^2$and five in less than 100,000 km^2. The gypsum salt deposits of the Jialingjiang Formation are mainly located in the eastern Sichuan Basin, including Wanzhou, Zhongxian, Liangping, Dianjiang, Changshou, Yubei, Jiangjin, Hechuan and Yunyang. There are 3 known salt basins (Shilong at Jiangjin, Baziyan atYubei and Dianjiang) and 1-30 layers of gypsum and halite, which is one of the important salt-forming periods in China.

3. Samples and analysis

Ten samples of carbonate rock, gypsum salt and rock salt from the Jialingjiang Formation drilled in the Dianjiang Salt Basin were selected for porosity and permeability experiments. The reference standard of experimental method is GB/T 29172-2012.

4. Results

The porosity of samples from the Jialingjiang Formation a drilling-core in the Dianjiang area are 0.63%-2.56%. The permeability of the samples are 0.0006-0.0018^{-3} μm^2.

Table 1. The porosity of samples from the Jialingjiang Formation a drilling-core in the Dianjiang area are 0.63%-2.56%. The permeability of the samples are 0.0006-0.0018^{-3} μm^2.

No.	Sample.No.	Sample state	Effective porosity %	Permeability×10^{-3} μm^2
1	Y230982001	cylindrical	1.69	0.001
2	Y230982002	cylindrical	2.56	0.0007
3	Y230982003	cylindrical	1.63	0.0007
4	Y230982004	cylindrical	0.91	0.0013
5	Y230982005	cylindrical	1.44	0.0013
6	Y230982006	cylindrical	1.23	0.0018
7	Y230982007	cylindrical	0.63	0.0011
8	Y230982008	cylindrical	2.08	0.001
9	Y230982009	cylindrical	1.28	0.0009
10	Y230982010	cylindrical	1.48	0.0006

5. Evaluation of sealing conditions of the high-steep fold belt in eastern Sichuan

5.1 CO_2 geological storage in the deep part of eastern Sichuan.

Carbon sequestration in carbonate rocks. The CO_2 in eastern Sichuan is sealed in structure or geology and pores. Structural or stratigraphic sealing mainly means that the upper caprock or reservoir blocks the upwards passage of CO_2 and prevents CO_2 from escaping. Pore sealing mainly means that CO_2 displaces the existing fluid to occupy and fill the pores in the rock, and the capture provided by capillary force can leave CO_2 in the pores of the reservoir structure. There are many gas fields in Chongqing, such as the Wolonghe gas field. The gypsum and salt deposits in the Jialingjiang Formation are generally divided into four sections, of which the first and third sections are mainly limestone and dolomitic limestone. The second and fourth members are dolomite and salt-dissolved breccia (gypsum-salt layer). When the fifth members of the Jialingjiang Formation and Leikoupo Formation remain, they may still be important gypsum and salt horizons. The formation temperature of the Triassic Jialingjiang Formation in the Sichuan Basin is relatively stable, basically maintained at 3℃/100 m; that is, the burial depth reaches 3000 m, which is the key depth for brittle-plastic transformation of gypsum rock of Jialingjiang Formation in the Sichuan Basin. Through the study of the burial history of the Sichuan Basin, it is found that during the Cretaceous sedimentary period, the burial thickness of Triassic strata in the Sichuan Basin reached approximately 3000 m, and the formation temperature reached 100°C. Gypsum rock has changed from brittle to plastic, sealing faults and fractures in the formation and improving sealing performance, which provides favorable conditions for carbon dioxide storage in Triassic carbonate-evaporite strata in the study area.

After CO_2 is injected into the ground, a phase equilibrium reaction occurs, which reduces the pH value of the formation water (Bachu et al.,2003). The diffusion coefficient of supercritical carbon dioxide is much higher than that of liquid carbon dioxide. When the pressure is lower than the critical pressure, the diffusion coefficient of carbon dioxide decreases rapidly with increasing pressure. When the pressure is higher, the pressure has relatively little effect on the diffusion coefficient of carbon dioxide. This will change the balance between the original salt water and rocks in the formation, leading to mineral dissolution or precipitation in the formation. The reaction of minerals in the brackish water layer of carbonate rocks is relatively simple; calcite, dolomite and other carbonate minerals are mainly dissolved, and secondary carbonate minerals such as ankerite and siderite are difficult to generate because there are no silicon-aluminum minerals to provide related ions such as Ca^{2+}, Fe^{2+} and Mg^{2+}(Cinar et al.2007).

Salt cavity is a potential carbon sequestration geologic body. The abandoned solution cavity formed by brine mining is an available underground space resource . Salt rock is a mineral aggregate composed of rock salt, and its main component is NaCl, which is a soluble evaporite minerals. The strata of the Jialingjiang Formation and Leikoupo Formation contain a large set of salt rocks with compact structure, low porosity and low permeability, which have good rheological properties. Salt rock is mainly mined by water solution and has self-healing characteristics (Liu et al.,2016), so it is recognized by scholars as the most suitable geological body for underground CO_2 geological storage(Kopp et al.,2009). The waste salt cavity may contain brine or salt water, which can be used as a good medium for CO_2 storage (Kumar et al.,2005). According to the achievements of the UK-China Prosperity Strategic Fund project "Geological Utilization and Storage Potential of Carbon Dioxide in Sichuan Basin and Early Demonstration Opportunities", the storage potential of Chongqing's CO_2 saline aquifer is 745 million tons. By the end of 2018, 11 rock salt mines had been discovered in the Chongqing city, including 7 rock salt mines and 4 bitterns. At present, the main salt mines in use are the Changshou Shuanglong Salt Mine and Hechuan Dashi Salt Mine, and the rest of the abandoned salt mines can be used as the objects of prestorage selection and evaluation.

5.2 Influence of faults on CO_2 storage

The basic geological conditions, such as porosity and permeability, of geological bodies around salt caves should be evaluated in the assessment of carbon dioxide storage in Chongqing. When CO_2 is injected into the fault, the pore pressure near it will increase, which will lead to the revival of the fault and the destruction of the caprock. The leakage of CO_2 along faults/fractures is mainly affected by the fracture opening and effective permeability. CO_2 leaks and migrates back to the atmosphere through faults, which provide potential leakage channels, leading to the pollution of the aquifer above the caprock. When a CO_2 salt cavern is sealed and located, it is necessary to determine the distribution of faults in this area by geophysical means and try to avoid unfavorable factors in the cavity-building area.

5.3 Effect of temperature on CO_2 storage

Study on the influence of temperature on geological storage of CO_2, including formation temperature and injection temperature. CO_2 becomes a supercritical fluid in the deep stratum, and its viscosity is only approximately one order of magnitude higher than that of atmospheric gas and much lower than that of liquid. Supercritical CO_2 has excellent mass transfer capacity. Higher temperatures will greatly reduce the fluid density, thus increasing the buoyancy in CO_2. The temperature change has a significant effect on the solubility of CO_2 in deep saline aquifers. Higher temperature leads to a great decrease in fluid density, which increases the buoyancy of CO_2, reduces the viscosity and increases the fluidity of CO_2. With increasing temperature, the diffusion coefficient of carbon dioxide also increases. With increasing pressure, the viscosity of carbon dioxide increases.

6. Conclusion

Carrying out carbon sequestration in Chongqing is conducive to the foundation of subsequent carbon neutralization, and geology comes first. CO_2 geological storage is the most direct and reliable method that can effectively reduce emissions. It is urgent to carry out scientific exploration on the influence of complex factors on the efficiency and safety of CO_2 geological storage. The factors that affect the storage capacity of CO_2 are the stratum structure (stratum dip), faults, lithology, thickness and porosity and permeability characteristics of the rock mass, salinity, temperature and pressure of deep groundwater, and injection temperature, which can be controlled by the outside world. Based on the comprehensive analysis of the regional geological background and the evolution characteristics of lithofacies and palaeogeography in the basin, the CO_2 covering the eastern Sichuan Basin was constructed.Abandoned salt caverns will be the primary underground storage space of choice for carbon sequestration in Chongqing

Acknowledgements

This paper is supported by the Chongqing Performance Incentive Project (Granted No. cstc2021jxjl900003), General Program of Chongqing Natural Science Foundation, China (Granted No. CSTB2022NSCQ-MSX0570) and Natural Science Foundation (Grant No.41902092).

References

1. Bachu S, Adams J J,2003. Sequestraton of CO2 in geological media in respose to climate change: capacity of deep saline aquifers to sequester CO2 in solution. Energy Conversion and Management 44(20), 3151-3175.
2. Bachu S, Bonijoly D, Bradshaw J, et al.2007. CO2 storage capacity estimaiton: methodology and gaps.

International Journal of Greenhouse Gas Control.1(4): 430-443.

3. Cinar Y, Riaz A, and Tchelepi H A. 2007.Experimental study of CO2 injection into saline formations. Paper SPE 110628 presented at SPE Annual Technical Conference and Exhibition, Anaheim, California, USA.
4. Kumar A, Ozah R. Noh M, et al. 2005.Reservoir simulation of CO2 storage in deep saline aquifers.Paper Spe 89343 (First presented at SPE/DOE Symposium on Improved Oil Recovery, Tulsa,Oklahoma, Apr. 17-21,2004).
5. Kopp A , Class H , Helmig R. 2009. Investigation on CO2 storage capacity in saline aquifers-Parts 1: Dimensional analysis of flow processes and reservoir characteristics. Internation Journal of Greenhouse Gas Control 3,263-276.
6. Liu W, Chen J, Jiang D Y, et al. Tightness and suitability evaluation of abandoned salt cacerns served as hydrocarbon energies storage under adverse geological conditions (AGC). Applied Energy,2016,178:708-720.

Elements characteristics and ecological risks of farmland soil with typical heavy metal high background in Chongqing

Jinhua Luan[1,2], Ruigang Zhang[1,2,*], Heming Tian[1,2], Jiahui Duan [3], Mohuan Ning [3], Lingling Mao [1,2], Feng Hu [1,2], Haitong Zhao [1,2], Changchao Li [1,2], Feng Liu [1,2], Sen Zhang [1,2]

[1] Chongqing Key Laboratory of Major Geological Events and Effects of resources and the environment (Chongqing Institute of Geology and Mineral Resources), 401120, China
[2] Observation and Research Station of Ecological Restoration for Chongqing Typical Mining Areas, Ministry of Natural Resources (Chongqing Institute of Geology and Mineral Resources), 401120, China
[3] Chongqing Institute of geological survey, Chongqing 401122, China

Abstract. To investigate the characteristics of elements in selenium and cadmium rich soil and the ecological risk of local agricultural products in the typical heavy metal high background area of Chongqing,Taking the Se and Cd enriched soil in study area as a research object, The effects of Cd and Se on agricultural products were studied by collecting and analyzing the element contents and available contents of soil and agricultural products samples, The results showed that The pH value of soil in the study area is 6.89 ~ 8.22, which is alkaline and rich in organic matter; The study area belongs to the typical high background area of heavy metals. The content of Cd in the soil is 0.52 ~ 1.84 mg·kg-1, the average value is 1.26 mg·kg-1, the content of se is 0.41 ~ 0.77 mg·kg-1, the average value is 0.64 mg·kg-1, and its element characteristics are rich in selenium and cadmium; The activation rates of Cd and Se were 6.63 ~ 28.26% and 0.17 ~ 1.58% respectively; The concentration of Cd and Se in the soil of the study area may be mainly from the parent material of soil, and it is closely related to the content of organic matter in the soil. The local agricultural products are less affected by the high background value of heavy metals in the soil, and there is no excessive Cd in the agricultural products, among which potatoes and sweet potatoes have reached the standard of selenium enrichment. The research suggests that the soil rich in selenium and cadmium in the study area may come from the soil parent material, which is caused by the natural geological background and closely related to the content of organic matter in the soil. The activation rate of Cd and Se in the soil is low, which has little impact on the ecological risk of local agricultural products. This paper has important guiding significance for the development and utilization of selenium-rich soil resources in typical heavy metal high background areas.

Key Words: Heavy metal; soil; elements characteristics; ecological risk; Chongqing

1. Introduction

At present, soil heavy metal pollution is an important environmental problem in China.Regional geological high background is an important factor affecting the enrichment of heavy metals in soil. It is reported that the enrichment of heavy metals in soil in some areas of Zhejiang, Pearl River Delta and southwest carbonate rock areas in China is affected by geological high background to varying degrees[1-2],Among them, the southwest region is a typical geological high background area, and the environmental problems caused by the natural enrichment of heavy metals such as cadmium in large areas of soil have attracted much attention[3-6].In addition to the obvious enrichment of heavy metals in the soil of heavy metal high background area, the soil is often accompanied by the enrichment of selenium. For example, the selenium enrichment area in the surface soil of the western region of Fuling is mostly concentrated in the heavy metal high background area[7].At present, many studies focus on heavy metal elements and ignore the health element selenium. In general, the biological activity of heavy metals in the high background area of carbonate rocks is low, and the risk of pollution to agricultural products is low[8].

In this paper, the geochemical characteristics and bioavailability of soil elements in typical heavy metal high background area of Chongqing were studied,At the same time, the edible parts of rice, potato, sweet potato, soybean, pepper and other crops planted in this area were sampled and tested, and the characteristics and ecological risks of selenium-rich and cadmium-rich soil elements in the study area were preliminarily discussed.The study found that heavy metals in the soil of the high background

* Corresponding author: sportagang@163.com

https://doi.org/10.1051/e3sconf/202447801024

area were significantly enriched, but the biological activity was low, and the risk of pollution to agricultural products was small. At the same time, selenium was enriched in the soil, and some selenium-rich agricultural products were found. This paper also provides a possible and new idea for the development and utilization of selenium-rich land resources in heavy metal high background areas.

Selenium is recognized as an essential nutrient and health element[1],The biological functions of selenium include anti-oxidation, anti-aging, enhancing immunity, preventing tumor, preventing Keshan disease and Kashin-Beck disease. However, there is a certain association between Se and Cd in soil[2],Cadmium is the most important pollutant in farmland soil in China, which has strong chemical activity and easy transformation. It has strong toxicity and carcinogenic effect. Therefore, the development and utilization of selenium-rich and cadmium-rich soil has been the focus of research at home and abroad[3],

In this paper, the characteristics of selenium and heavy metals and their bioavailability in selenium-rich and cadmium-rich soils were studied in typical farmland with high background of heavy metals in Chongqing. Meanwhile, the edible parts of rice, potato, sweet potato, soybean and pepper were sampled and analyzed, and the element characteristics and ecological risk of selenium-rich and cadmium-rich soils in the study area were discussed.

2. Materials and Methods

2.1 Overview of the study area

The study area is located in Changshou District, Chongqing (Fig. 1), in the east of Sichuan Basin, and the geological structure belongs to the northwest wing of Mingyuexia anticline. The study area is located in the Jialingjiang Formation of Lower Triassic of Mesozoic Erathem, and the parent rock of soil formation is mainly limestone and bioclastic limestone.The main soil type in the study area is paddy soil.The exposed area of carbonate rocks in the world accounts for about 12% of the land area[4],The background value of soil cadmium in karst carbonate rock area in Southwest China is much higher than the national average value. According to He Shaolin et al. [9], the geochemical background value of cadmium in soil and sediment in Guizhou Province is 0.31mg·kg-1, which is 2.5~3.5 times of that of China.The National Soil Pollution Survey Bulletin published in 2014 [10] shows that the over-standard rate of cadmium in cultivated soil in China is 7%, and the problem of cadmium pollution in soil in Southwest China is relatively serious, and the high geological background is one of the important reasons.Studies have shown that. The high background of cadmium in geochemical anomalies is common in soils in sedimentary rock development areas represented by carbonate rocks[11-13],The study area is located in carbonate rock deposition area, which belongs to typical heavy metal Cd high background area.

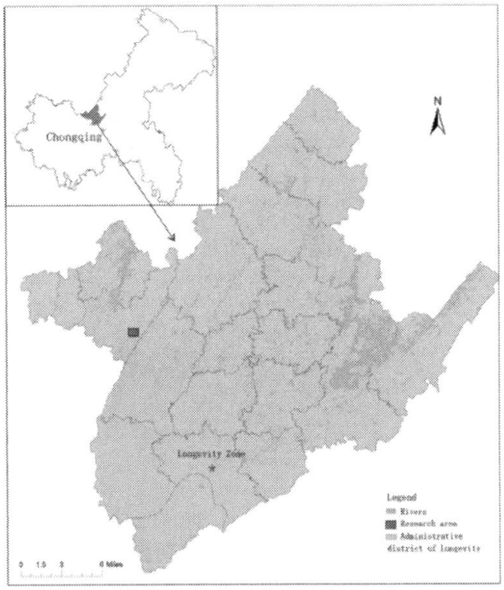

Fig. 1 Map showing location of study area

2.2 Sample Collection and Testing

On the basis of 1: 50,000 land quality geological survey data in Changshou District, the continuous farmland area with soil cadmium (Cd) content greater than 0.3 mg·kg-1 and soil selenium (Se) content greater than 0.4 mg·kg-1 was selected as the study area. Twelve surface soil samples (Fig. 2) were collected this time, with the sampling depth of 0~20cm. The sampling pit was dug with a shovel, and the soil in contact with the shovel was removed with a wooden shovel. Then the soil samples were collected from top to bottom. During sampling, the main sampling pit was taken as the center, and 4 sampling points were selected within 30m nearby, and equal parts were combined into a mixed sample.Air dry the soil samples at room temperature, crush them with plastic rods, remove sand and roots, etc., pass all the samples through a nylon sieve with a 2mm aperture, mix them thoroughly, and then take 200g of samples for analysis and testing by quartering method, and the remaining samples are reserved.Collect edible parts of local bulk crops, including 2 pieces of rice, 2 pieces of potato, 2 pieces of sweet potato, 2 pieces of soybean and 1 piece of pepper.

The heavy metal elements Cr, Cu and Zn in the samples were determined by X-ray fluorescence spectrometry (XRF, Panaco PW4400, Netherlands).heavy metal element Cd, Ni and Pb were determined by inductively couple plasma mass spectrometry (ICP-MS, United States thermoelectricity Xseries II);As, Hg and Se in the samples were determined by atomic fluorescence spectrometry (AFS, Beijing Haiguang AFS-9760).Corg in soil was determined by redox volumetric method (VOL).The pH value of soil was determined by ion selective electrode method (ISE, PHSJ-3F).The available Cd, Zn and Pb in soil were determined by inductively coupled plasma mass spectrometry (ICP-MS), and the available Se was determined by atomic fluorescence spectrometry (AFS).During the test, the national standard soil reference substance is added for quality control, and the soil and plant sample elements are tested by Chongqing Mineral Resources Supervision and Testing Center of Ministry of Land and Resources.

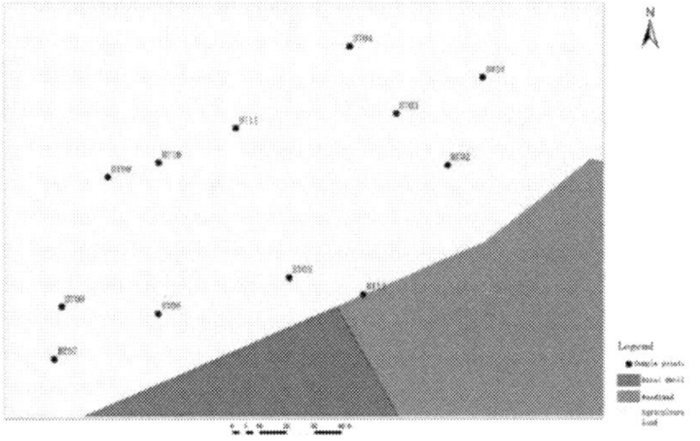

Fig.2 Location of study area and sampling sites

3. Results and Discussion

3.1 Characteristics of soil pH and organic matter

It can be seen from Table 1 that the soil pH value in the study area is between 6.89 and 8.22, with an average value of 7.77. According to the soil pH value classification standard [14] of the Land Quality Geochemical Evaluation Specification (DZ/T0295-2016), the overall performance is alkaline.The content of organic matter in the soil was 22.41~34.45 g·kg-1, with an average of 33.52 g·kg-1. The content of organic matter in the soil reached the second class (relatively rich).in comparison with that average value of pH (6.41) and organic matter (15.17 g·kg-1) in the surface soil of 1:50000 land quality survey in Changshou District, the soil pH value in the study area was high and the soil was alkaline.The content of organic matter was more than 2 times of the average background value of surface soil organic matter in Changshou District, and the content of soil organic matter was more abundant.

Table 1 Statistical table of soil pH value and organic matter content

Item	Sample No.												Statistics		
	BT01	BT02	BT03	BT04	BT05	BT06	BT07	BT08	BT09	BT10	BT11	BT12	Mean	Min	Max
pH	7.89	7.61	8.14	7.98	7.47	7.53	7.38	7.91	8.05	8.20	8.22	6.89	7.77	6.89	8.22
organic matter (g·kg^{-1})	34.48	37.76	37.76	33.45	38.45	33.96	33.96	28.27	30.34	36.55	34.82	22.41	33.52	22.41	38.45

3.2 Soil element characteristics

3.2.1 Element content characteristics

According to the statistical analysis results (Table 2), the average contents of heavy metals in the soil of the study area are 15.95, 1.26, 91.35, 0.13, 32.58, 63.28, 53.89 and 141.53 mg·kg-1, respectively.Compared with the background values of As (4.54), Cd (0.26), Cr (72.75), Hg (0.05), Pb (27.55), Ni (30.93), Cu (23.27) and Zn (80.87) in the surface soil of Changshou District, the contents of heavy metals in the soil of study area were 1.18~4.8 times higher, and the content of Cd was 4.8 times higher. It is reported that the background value of Cd content in soil in China is 0.1 mg·kg^{-1}, among which the background value of Cd content in main agricultural soil in China ranges from 0.01 to 1.34 mg·kg^{-1}, with an average value of 0.12 mg·kg^{-1}[15]. The soil in the study area belongs to typical heavy metal high background area. The content of Cd in the soil ranged from 0.52 mg·kg-1 to 1.84 mg·kg-1, exceeding the risk control standard of soil pollution in agricultural land(GB15618-2018), among which 5 samples were slightly polluted, 5 samples were slightly polluted, 2 samples were moderately polluted, and other heavy metals in the soil did not exceed the standard. According to the classification standard of soil nutrient grade in the Specification for Geochemical Evaluation of Land Quality, the contents of Cu and Zn in the soil reached the rich level.

The content of Se in the soil of the study area ranged from 0.41 to 0.77 mg·kg^{-1}, with an average of 0.64 mg·kg^{-1}. According to the standard of Se-rich soil (> 0.4 mg·kg^{-1}) of the Geochemical Evaluation of Land Quality, the soil of the study area is Se-rich soil.The content of Mn in the soil ranges from 244 to 1702 mg·kg^{-1}, with an average of 894.17 mg·kg^{-1}, and the content of Fe in the soil ranges from 3.74 to 6.06%, with an average of 4.88%. According to the classification standard of soil nutrient in the Geochemical Evaluation of Land Quality, the content of Mn in the soil of the study area is rich (>700 mg·kg^{-1}), and the content of Fe in the soil is rich (>3.71%).

https://doi.org/10.1051/e3sconf/202447801024

Table 2 Statistics of heavy metal content in soil (mg·kg-1)

Item	Sample No.												Statistics		
	BT01	BT02	BT03	BT04	BT05	BT06	BT07	BT08	BT09	BT10	BT11	BT12	Mean	Min	Max
As	14.60	14.30	13.20	12.40	16.00	21.10	17.80	17.40	17.30	15.00	11.90	20.40	15.95	11.90	21.10
Cd	1.40	1.84	1.81	1.51	1.84	1.11	0.69	0.67	0.92	1.32	1.44	0.52	1.26	0.52	1.84
Cr	95.48	99.91	86.25	87.71	88.90	101.00	88.70	94.49	81.78	87.85	85.97	98.14	91.35	81.78	101.00
Hg	0.14	0.14	0.12	0.12	0.14	0.15	0.15	0.13	0.13	0.12	0.12	0.13	0.13	0.12	0.15
Pb	27.80	32.20	25.10	28.30	35.60	38.00	39.20	33.30	36.20	29.30	27.80	38.20	32.58	25.10	39.20
Ni	69.50	96.80	82.40	63.00	78.80	64.50	48.20	44.20	46.80	63.40	65.70	36.10	63.28	36.10	96.80
Cu	62.20	69.10	67.30	50.20	64.20	74.60	41.90	39.50	35.40	49.90	57.10	35.30	53.89	35.30	74.60
Zn	150.00	194.00	175.00	131.00	187.00	153.00	114.00	107.00	112.00	138.00	142.00	95.30	141.53	95.30	194.00
Se	0.69	0.76	0.71	0.72	0.63	0.54	0.59	0.55	0.57	0.72	0.77	0.41	0.64	0.41	0.77
Mn	1415	428	1702	1150	996	353	244	707	714	1193	1094	734	894.17	244	1702
Fe (%)	6.06	5.63	5.77	4.95	5.02	5.02	4.05	4.17	3.74	4.96	4.98	4.19	4.88	3.74	6.06
Cd Pollution	Light Pollution	Light Pollution	Moderate Pollution	Light Pollution	Moderate Pollution	Slight Pollution	Slight Pollution	Slight Pollution	Slight Pollution	Light Pollution	Light Pollution	Slight Pollution	—	—	—

3.2.2 Available amounts of Cd, Se and Zn in soil

It can be seen from the statistical data in Table 3 that the content of available Cd in the soil of the study area ranges from 0.11 to 0.49 mg·kg^{-1}, with an average value of 0.25 mg·kg^{-1}. The content of available Cd accounts for 6.63 to 28.26%, with an average value of 18.91%. It can be seen from Figure 3 that the available amount of Cd in the soil is proportional to the total amount.

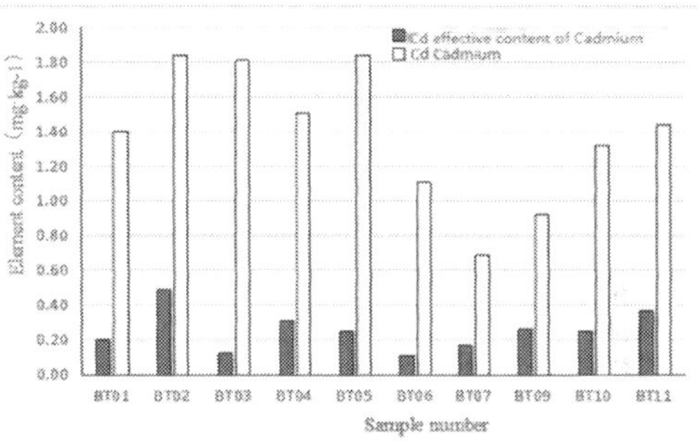

Fig.3 Cd content of soil in study area

The content of available Se in soil ranged from 0.001 to 0.011 mg·kg^{-1}, with an average of 0.0072 mg·kg^{-1}. The content of available Se in soil ranged from 0.17 to 1.58%, with an average of 1.06%.The content of available Zn in the soil ranged from 2.75 to 9.86 mg·kg^{-1}, with an average of 5.05 mg·kg^{-1}. The content of available Zn accounted for 1.83 to 5.08%, with an average of 3.29%.It can be seen that the content of available Cd in soil is relatively high, and the activity is large. Although the content of Se and Zn in soil is rich, the content of available Cd is relatively low.

Table 3 Statistics of Cd、Se and Zn available content in soil

Item	Sample No.										Statistics		
	BT01	BT02	BT03	BT04	BT05	BT06	BT07	BT09	BT10	BT11	Mean	Min	Max
Cd Effective dose (mg·kg^{-1})	0.20	0.49	0.12	0.31	0.25	0.11	0.17	0.26	0.25	0.37	0.25	0.11	0.49
Cd Proportion of effective amount (%)	14.29	26.63	6.63	20.53	13.59	9.91	24.64	28.26	18.94	25.69	18.91	6.63	28.26
Se Effective dose (mg·kg^{-1})	0.008 0	0.008 0	0.007 0	0.009 0	0.007 0	0.004 0	0.001 0	0.009 0	0.008 0	0.011 0	0.007 2	0.001 0	0.011 0
Se Proportion of effective amount (%)	1.16	1.05	0.99	1.25	1.11	0.74	0.17	1.58	1.11	1.43	1.06	0.17	1.58
Zn Effective dose (mg·kg^{-1})	2.75	9.86	4.60	3.86	7.92	6.10	3.39	3.48	4.06	4.50	5.05	2.75	9.86
Zn Proportion of effective amount (%)	1.83	5.08	2.63	2.95	4.24	3.99	2.97	3.11	2.94	3.17	3.29	1.83	5.08

3.2.3 Correlation analysis of soil elements

The Pearson correlation coefficient (Table 4) of As, Cd, Cr, Hg, Pb, Ni, Zn, Cu, Mn and Fe in the topsoil of the study area and the soil pH value and organic matter was analyzed by SPSS19.0 software. The data showed that the available amount of Cd, Se and Zn in the soil was positively correlated with their total amount, among which the available amount of Se and Zn was significantly positively correlated with the total amount ($p<0.05$), and the correlation coefficients were 0.634 and 0.71 respectively.The effective amount of Se is significantly positively correlated with soil pH value (Figure 4), and negatively correlated with organic matter to a certain extent, indicating that the high content of organic matter in soil has adsorption and complexation effect on selenium in soil, which inhibits the absorption of selenium by crops to a certain extent.Soil organic matter is highly significantly positively correlated with total soil Cd and Zn ($p< 0.01$), and significantly positively correlated with total Ni, indicating that organic matter in soil is closely related to Cd, Zn and Ni (Figure 5).

The correlation coefficient between heavy metal elements can indicate the degree of similarity of their source pathways. Generally, heavy metals with higher correlation coefficient have dependence relationship and may have similar source pathways [16].The correlation coefficients of As-Hg, As-Pb, Cd-Zn, Cd-Ni, Ni-Cu and Ni-Zn in soil were higher and showed very significant positive correlation, and the correlation coefficients of Cd, Zn and Ni were very significant positive correlation, the correlation coefficients were 0.956, 0.928 and 0.867 respectively, which indicated that Cd, Zn and Ni in soil might have similar sources.

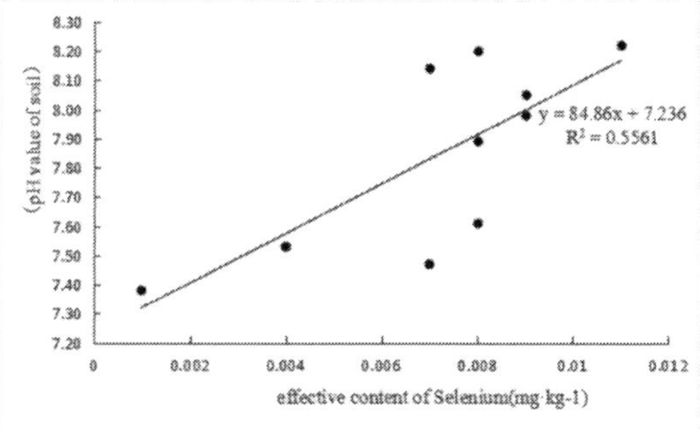

Fig.4 Trend chart of available selenium content and pH value in soil

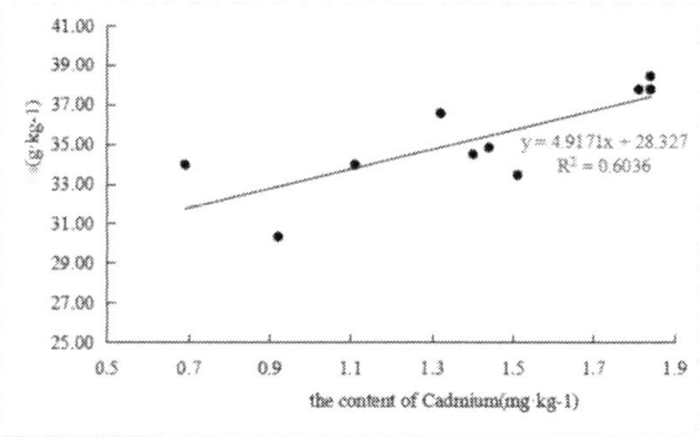

Fig.5 the trend map of cadmium content and organic matter in Soil

The correlation coefficient between Se and Cd was 0.671, and the correlation between Se and Cd was negative or not.The correlation coefficient between Mn and Cd, Se, Zn, Ni in the soil was positive, and the correlation coefficient between Mn and Cd, Se was higher, while the correlation coefficient between Fe and Cd, Ni, Zn, Se was significantly positive, and the correlation coefficients were 0.758, 0.748, 0.785, 0.735 respectively. It can be seen that Cd, Se, Zn, Ni in the soil of the study area were closely related to Mn, Fe, and the existing forms of Cd and Se in the soil might be related to Fe and Mn, which led to the low content of their available forms. It is generally believed that Ni is mainly affected by geochemical genesis [17], Fe and Mn are important substances of soil-forming parent material, among which Mn is regarded as a characteristic element of natural source and a reference element used to determine the impact of human activities on heavy metal accumulation [18]. It is inferred that Cd, Se, Zn and other elements in the soil of the study area may be mainly derived from soil-forming parent material, and their high content may be mainly caused by natural geological background.

3.2.4 *Analysis on the Sources of Cd and Se in Soil*

Some studies have shown that the soil in the sedimentary rock development area represented by carbonate rock generally has high background phenomenon of geochemical anomaly of cadmium, and the background value of heavy metal cadmium in the soil type statistical unit is the largest (1.115 $mg \cdot kg^{-1}$), the soil cadmium content in karst area of southwest China is much higher than the national average value, and the soil cadmium in carbonate rock area has the characteristics of high background value and large distribution area [19].According to Baize, the excessive accumulation of cadmium in soil mainly comes from soil-forming parent rocks, which were formed in oolitic limestone in ancient shallow water environment, with cadmium content up to 8.15 mg·kg-1[20]。The stratum of the study area is Jialingjiang Formation of Lower Triassic, and the parent rock is mainly limestone and bioclastic limestone, which belongs to typical carbonate sedimentary area. Therefore, Cd in the soil of the study area may come from the parent rock (limestone).

According to Xiang Jiqian et al., the selenium in the soil of high selenium area in Enshi Prefecture mainly comes from black rock series and coal measures strata, especially siliceous shale and carbonaceous shale with high selenium content [21].Generally, the proportion of selenium enriched in carbonaceous shale is relatively large, and the selenium content of soil developed from shale is the highest [22].Although the lithology of the study area is mainly limestone, the lithology of the underlying Feixianguan Formation is mainly shale and mudstone, while the overlying Xujiahe Formation is sandwiched with thin coal seam or carbonaceous shale. The study area is located in low-lying terrain with mountains on both sides. It is speculated that the selenium in the soil may come from the surrounding selenium-rich black rock series (coal seam or carbonaceous shale).

Soil organic matter and selenium have complex interaction forms, which have important influence on the adsorption of soil selenium. Generally speaking, the soil with high organic matter content has higher selenium content [23]. The contents of Se and organic matter in the soil of the study area are both high, so the organic matter in the soil may have strong adsorption effect on Se, so that the Se transported nearby is adsorbed and enriched in the soil.This is also consistent with the results of correlation analysis of various elements in the soil that Cd and Se may be mainly derived from soil parent materials.

Table4 Correlation coefficients between elements in soil

Element	Effective Cd	Effective Se	Effective Zn	pH	Organic matter	As	Cd	Cr	Hg	Pb	Ni	Cu	Zn	Se	Mn	Fe
Effective Cd	1															
Effective Se	0.579	1														
Effective Zn	0.492	0.011	1													
pH	0.118	0.746*	-0.466	1												
Organic matter	0.151	-0.005	0.622	-0.142	1											
As	-0.517	-0.704*	0.073	-0.623	-0.306	1										
Cd	0.295	0.438	0.586	0.098	0.786**	-0.530	1									
Cr	0.067	-0.311	0.524	-0.544	0.275	0.367	0.186	1								
Hg	-0.255	-0.765**	0.266	-0.866**	-0.004	0.805**	-0.326	0.635*	1							
Pb	-0.163	-0.642*	0.187	-0.627	-0.109	0.781**	-0.553	0.205	0.742*	1						
Ni	0.341	0.156	0.744*	-0.224	0.712*	-0.311	0.867**	0.515	-0.008	-0.418	1					
Cu	-0.093	-0.115	0.594	-0.391	0.464	0.139	0.625	0.708*	0.270	-0.218	0.822**	1				
Zn	0.265	0.258	0.710*	-0.056	0.866**	-0.368	0.956**	0.379	-0.123	-0.397	0.928**	0.732*	1			
Se	0.551	0.634*	0.140	0.453	0.476	-0.886**	0.671*	0.068	-0.542	-0.759*	0.573	0.188	0.599	1		
Mn	-0.157	0.554	-0.351	0.708*	0.285	-0.676*	0.529	-0.391	-0.714*	-0.779**	0.156	0.008	0.352	0.594	1	
Fe	0.071	0.251	0.274	0.130	0.636*	-0.419	0.758*	0.521	-0.122	-0.586	0.748*	0.671*	0.785**	0.735*	0.543	1

**Extremely significant correlation（P<0.01）; *Significant correlation（P<0.05）.

3.3 Ecological risk of agricultural products

Table 5 shows the results of Cd, Se and Zn content in agricultural products in the study area. Cd in potato and red potato was not detected in agricultural products in the study area. Cd content in rice was 0.021~0.026 mg·kg^{-1}, Cd content in pepper was 0.019 mg·kg^{-1}, and Cd content in yellow bean was 0.078~0.089 mg·kg^{-1}. According to the national food safety standard,The results showed that the content of Cd in rice was 0.2 mg·kg^{-1} according to the limited standard of paddy, 0.05 mg·kg^{-1} according to the limited standard of fresh vegetables and 0.2 mg·kg^{-1} according to the limited standard of beans.

The content of Se in potato and sweet potato in the study area is 0.022~0.023 mg·kg^{-1} and 0.019~0.021 mg·kg^{-1} respectively. According to the standard of selenium-rich potato (0.015~0.15 mg·kg^{-1}), potato and sweet potato in the study area meet the standard of selenium-rich.The Se content in rice is 0.013~0.015 mg·kg^{-1}, which does not reach the standard of selenium-rich rice (0.04~0.3 mg·kg^{-1}).The content of Se in pepper and soybean is similar to that in other agricultural products.In general, the ecological risk of crops in the study area is low, and there is a certain selenium-rich potential.

According to the statistical data of China food composition of Yang Yuexin et al., the national average Zn content in soybean is 33.4 mg·kg-1, and the average Zn content in rice is 17 mg·kg^{-1}. The Zn content in agricultural products in the study area is generally low, but the Zn content in soybean is obviously rich (68.1~89.3 mg·kg^{-1}), and the Zn content in rice is slightly higher than the national average level (17.9~21.4 mg·kg^{-1}).

Table 5 Results of element content in agricultural products (mg·kg^{-1})

Sample No.	Sample Type	Cd	Se	Zn
TDZW01	Potato	<0.01	0.023	3.74
TDZW02		<0.01	0.022	2.44
LJZW01	Capsicum	0.019	0.02	2.77
HSZW01	Sweet potato	<0.01	0.019	1.42
HSZW02		<0.01	0.021	1.49
HDZW01	Soybean	0.089	0.028	89.3
HDZW02		0.078	0.014	68.1
SDZW01	Rice	0.026	0.013	17.9
SDZW02		0.021	0.015	21.4

4. Conclusions

The soil in the study area is alkaline, pH 6.89~8.22, average 7.77;The content of organic matter in the soil is rich, ranging from 22.41 to 34.45 g·kg^{-1}, with an average of 33.52 g·kg^{-1}, which is more than twice the background average of organic matter in the surface soil of Changshou District.

The correlation coefficients of As-Hg, As-Pb, Cd-Zn, Cd-Ni, Ni-Cu and Ni-Zn in the soil were higher and showed extremely significant positive correlation, and the correlation coefficients of Cd, Zn and Ni were extremely significant positive correlation, which indicated that Cd, Zn and Ni in the soil might have similar sources. The study area was a typical carbonate sedimentary area, and Cd in the soil might come from the soil parent rock(limestone), organic matter Se in soil is closely related, Cd and Se may be mainly derived from soil parent materials.

The local agricultural products were less affected by the high background value of heavy metals in soil, and there was no excessive Cd in agricultural products, so the ecological risk was low. Among them, potato and sweet potato reached the standard of selenium-rich food and had certain selenium-rich potential.

In this study, there are many kinds of agricultural products in the study area, but the number of samples is small. In the future research, the number of samples should be increased. It is suggested to strengthen the study on the migration and enrichment of heavy metals and selenium in soil-plants. With the continuous improvement of people 's demand for healthy food, the existing natural selenium-rich soil resources can not meet the urgent need for the scientific development and utilization of selenium-rich land resources in high background areas of heavy metals. This study can provide a rich scientific basis for this demand.

Acknowledgments

This study was supported by Chongqing planning and Natural Resources Bureau Projects "Four 1:50,000 land quality geological survey sheets (Changshou, Shizitan and others) in Chongqing city (Chongqing Bureau of Planning and Natural Resources [2022] No.61)"and "Chongqing Scientific Research Institutions Performance Incentive Guidance Project (No. cstc2021jxj190003)"

References

1. Xia X Q,JiJF,ZhangC S,et al.Carbonate bedrock control of soil Cd background in Southwestern China:its extent and influencing factors based on spatial analysis. Chemosphere, 2022, 290, doi:10.1016/j. chemosphere. 2021.133390.
2. Chen T Y, Zhao H F, Wu K N, et al. Distributional characteristics and source identification of cadmium in soils of the Pearl River Delta, China. Bulletin of Environmental Contamination and Toxicology, 2021,106(1):75-85.
3. Duan Y R,Yang ZF,Yu T,et al.Geogenic cadmium pollution in multi-medians caused by black shales in Luzhai,GuangxiEnvironmental Pollution,2020,260,doi:10.1016/j. envpol. 2019. 113905.
4. Yang Q,Yang ZF,Zhang QZ,et al.Transferability of heavy metal(loid)s from karstic soils with high geochemical background to peanutseeds. Environmental Pollution, 2022, 299, doi: 10.1016/j.envpol.2022.118819.
5. Liu Y Z,Xiao T F,Zhu Z J,et al.Geogenic pollution, fractionation and potential risks of Cd and Zn in soils from a mountainous region underlain by black shale. Science of the Total Environment, 2021,760, doi:10.1016/j.scitotenv. 2020.143426.
6. Liu Y Z, Chen Z J, Xiao T F, et al. Enrichment and environmental availability of cadmium in agricultural soils developed on Cd-rich black shale in southwestern China. Environmental Science_and Pollution Research,2022,29(24):36243-36254.
7. Wen Y B,Li W,Yang Z F,et al.Evaluation of variousapproaches to predict cadmium bioavailability to rice grown in soils with high geochemical background in the karst region, Southwestem China. Environmental Pollution, 2020. 258, doi:10.1016/j.envpol.2019.113645.
8. MAOLingling, TIANHeming, ZHANGXiong, etal. 2022. Geochemical characteristics and ecological effects of Se and Zn in topsoil in Western Fuling, Chongqing. E3S Web of Conferences, 352(03022) .
9. HE Shao-ling, LONG Chao -lin, LIU Ying-zhong, et al. Geochemistry and Environment of the Cadmium in the Soil and Sediments at the Surface of Guizhou Province. GuiZhou Geology, 2004, 21(04) : 245-253.
10. Ministry of environmental protection,Ministry of land and resources. The results of a national soil survey. Environmental Education, 2014(6): 8-10.
11. Lalor G C, R attray R , Simpson P, et al. Heavy metals in Jamaica. Part 3: the distribution of cadmium in Jamaican soils. R ev. Int. Contam.Ambient, 1998,14(1) :7-12.
12. LI Li-hui, WANG Bao-lu. Geochemical Characteristics of As and Cd in soils of YunNan Province. Geophysical and Geochemical Exploration, 2008, 32(5) : 497-501.
13. Quezada Hinojosa R P, Matera V, Adatte T, et al. Cadmium distribution in soils covering Jurassic oolitic limestone with high Cd contents in the Swiss Jura. Geoderma, 2009, 150 (3-4) :287 – 301.
14. Ministry of Land and Resources of PRC. Standard for geochemical evaluation of land quality DZ/T0295-2016 [S] . Beijing: Geological Press, 2016.
15. LUO Xu-qiang, WANG Shi-jie, LIU Xiu-ming.Geochemical characteristics of the soil cadmium in the process of karst rocky desertification. Ecology and Environmental Sciences, 2009, 18(1): 160 - 166.
16. Zheng Yuan-ming, Luo Jin-fa, Chen Tong-bin, et al. Cadmium accumulation in soils for different land uses in Beijing.Geographical Research, 2005, 24(4): 542-548.
17. Martin J. A. R. , Arias M. L. , Corbi J. M. G. . Heavy Metals Contentsin Agricultural Topsoil in the Ebro Basin (Spain). Application of theMultivariate Geostatistical Methods to Study Spatial Variations.Environmental Pollution , 2006 , 144 (3):1001 -1021.
18. Li X , Poon C , Liu P S . Heavy metal contamination of urban soils and street dusts in Hong Kong. Applied Geochemistry, 2001, 16(11):1361-1368.
19. LIU Hong-yan, TU Yu, GU Xiao-feng, et al. Accumulative Effect and Environmental Impact of Cadmium in Farmland Soil with High Background of Geochemistry. Journal of Mountain Agriculture and Biology, 2018, 37(05):5-9.
20. Baize D , Sterckeman T . Of the necessity of knowledge of the natural pedo-geochemical background content in the evaluation of the contamination of soils by trace elements. 2001, 264(1-2):0-139.
21. XIANG Ji-qian, ZHU Yun-fen, YIN Hong-qing. Current Status and Thoughts on the Development and Utilization of Selenium Resources in Enshi Autonomous Prefecture. Hubei Agricultural Sciences, 2016, 55(17): 4366-4369.
22. LI-Juan, LONG Jian, WANG Jing-Ren. Geochemical Characteristics of Selenium in Soils of Kaiyang Region , Guizhou Province. Chinese Journal of Soil Science, 2004(05):579-582.
23. SHAO Zong-yuan. The Migration of Selenium and Cadmium in Soil-Rice System in a Typical Selenium and Cadmium-Rich Area. Anhui:Anhui Normal University, 2017:20-21.

Research on Bidding Model of Electricity Selling Company Considering Green Certificate Trading and Electricity Selling Blocking

Shuo Yin [1,*], Yao Lu [1], Yongle Zheng [1], Xiangyu Jiao [2]

[1] State Grid Henan Economic Research Institute, Zhengzhou, China
[2] School of Mathematics and Statistics, Lanzhou University, Lanzhou, China

Abstract. This article considers the optimal bidding model of traditional thermal power generation companies' electricity sales companies, taking into account transmission congestion, under the influence of new policies such as the renewable portfolio standard (RPS) and the renewable exchange certificate (REC). Firstly, the transaction cost of green certificates is included in the cost of electricity sales, and an optimization model is established with the goal of maximizing the profit of the electricity sales company; Secondly, the node electricity price method is used to manage transmission congestion, and an optimization model is established with the goal of minimizing the cost of transmission congestion; Finally, multi agent technology and dynamic game theory are used to solve the optimal bidding strategy of the electricity selling company.

Keywords: Green Certificate Trading Renewable Energy Quota Transmission Blockage Bidding Model.

1. Introduction

With the gradual improvement of international and domestic standards for energy cleanliness and environmental friendliness, as well as the continuous progress and improvement of technology, China's energy development in the new era must adhere to the overall requirements of building a clean, low-carbon, safe and efficient energy system, and firmly follow the path of energy development with Chinese characteristics. Due to its renewable nature and good environmental friendliness, renewable energy has been regarded as the main means of solving energy shortages and achieving environmentally friendly development both domestically and internationally. By setting renewable energy quota targets, the proportion of renewable energy in the entire power system can be increased. This article first incorporates the cost of green card transactions into the cost of electricity sales, and establishes an optimization model with the goal of maximizing the profit of the electricity sales company; Secondly, the node electricity price method is used to manage transmission congestion, and an optimization model is established with the goal of minimizing the cost of transmission congestion; Finally, multi agent technology and dynamic game theory are used to solve the optimal bidding strategy of the electricity selling company.

2. Electricity trading mode under green certificate trading mechanism

The value of green certificate trading in the electricity trading model under the green certificate trading mechanism lies in affirming the driving role of the market's "invisible hands" in actively consuming renewable energy. Green certificate trading attempts to promote normal competition in the market by fully engaging market participants in the game, replacing government subsidies with market profits, and alleviating the pressure of high subsidies. Wind power enterprises can earn profits by selling green electricity from wind turbines, and can also earn excess profits by selling surplus green certificates.

Transactions in the renewable energy trading market can be divided into two categories:

The first category is that traditional energy generators need to complete their own renewable energy quota tasks generated by the necessary transactions. The traditional energy power generation enterprises' own renewable energy power capacity is insufficient, which is lower than the minimum quota requirements. It must subscribe to the green certificates of other renewable energy companies or other power companies with surplus green certificates.

The second category is that green certificates can be traded in a trading market for market participants, such as (renewable energy units, thermal power units, electricity consumers). The essence of market circulating goods lies

* Corresponding author: yinshuo1111@163.com

in market credit endorsement. Green certificates on the trading market include the actual use value and credit value of wind power as a commodity. After the government uses its credibility to give credit value to the green certificate, the green certificate has the use value and credit value while circulating among the market participants. Therefore, analogous to the securities market, the two parties involved in the transaction of green certificates must develop a reasonable contract signing mechanism. According to the different needs of users, the duration of signing contracts can be divided into long-term and short-term. Large power customers and thermal power units can sign a large number of medium - and long-term contracts, while ordinary power consumers can sign short-term contracts.

Figure 1 shows the relationship between participants in the green certificate trading market. It is assumed that the enterprises selling green certificates in the electricity market are wind power enterprises.

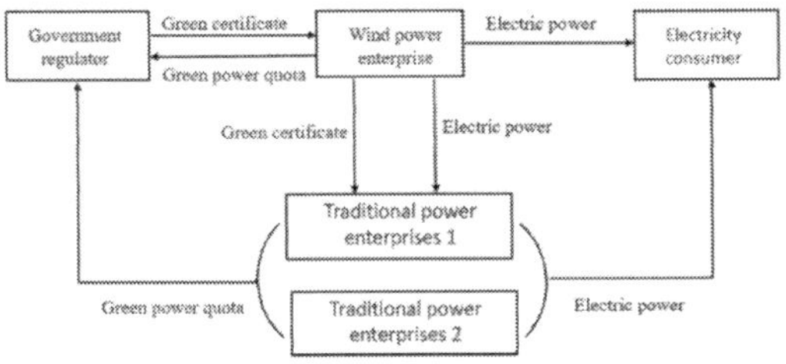

Figure 1. Wind power trading market reference chart.

3. A bidding model of electricity selling companies considering the impact of renewable energy quota and green certificate trading

For the convenience of description, only one trading period is considered in this paper, that is, the unit start-stop constraint is not considered. Suppose that M sales companies in the electricity market are set up by M thermal power producers, each of which has only one thermal power unit (but this method is also applicable to the general situation).

3.1 The cost function and quotation function of the selling company

The Since the renewable energy quota power generation of traditional thermal power generators is all generated by green certificate trading, the $i\left(i=\{1,2,\ldots,M\}\right)$ the green certificate transaction cost of electricity sales company is:

$$C_{ri}\left(P_i\right)=\rho\cdot kP_i \tag{1}$$

where ρ is the trading price of wind power green certificate, referring to the online trading situation of the green certificate subscription platform in July 2017, ρ is 15$/MWh; k is the proportion of renewable energy generation quota stipulated by the National Development and Reform Commission, meeting $k\geq 15\%$; P_i is the total amount of electricity sold per unit trading period.

The cost function of electricity sold by electricity selling company i is:

$$C_{ti}(P_{ti})=0.5a_iP_{ti}^2+b_iP_{ti}+c_i \tag{2}$$

Where, P_{ti} is the electricity sold per unit trading period, then $P_{ti} = \left(1-k\right) P_i$ is the power generation of thermal power units; a_i , b_i , c_i are the cost coefficient of electricity sold by the selling company.

After considering the transaction cost of green certificate, the cost function of electricity selling company i becomes:

$$\begin{aligned}C_i(P_i)&=C_{ti}(P_{ti})+C_{ri}(P_{ri})\\&=0.5a_i[(1-k)P_i]^2+b_i(1-k)P_i+c_i+\rho kP_i\\&=0.5a_i(1-k)^2P_i^2+[b_i(1-k)+\rho k]P_i+c_i\end{aligned} \tag{3}$$

the corresponding marginal cost curve is:

$$\begin{aligned}C_i(P_i)&=a_i(1-k)^2P_i+b_i(1-k)+\rho k\\&=\alpha_i+\beta_iP_i\end{aligned} \tag{4}$$

Where, α_i and β_i are the marginal cost parameters of the electricity selling company i.

The electricity selling company determines its own quotation based on load prediction, cost estimation, risk appetite, probability description of competitors' quotation behavior and other factors. Assuming that the market rules require the quotation function of the electricity selling company to be a linear function, in order to more directly reflect the characteristic that the actual quotation curve of the electricity selling company changes around the real marginal cost curve, the quotation function of the electricity selling company i is set as follows:

$$\xi_i(\lambda_i,P_i)=\lambda_i(\alpha_i+\beta_iP_i) \tag{5}$$

Where λ_i is the quotation parameter of the electricity selling company i.

3.2 Unconstrained bidding model of electricity selling company

After receiving the quotation of the selling company, the power market trading center first carries out unconstrained scheduling. In the case that the transmission capacity constraint is ignored, the real-time electricity price of all nodes in the system is the same, set to ξ_x. At this time, the electricity selling cost function of the electricity selling company is:

$$W_1=\sum_{i=1}^{M}\xi_xP_i \tag{6}$$

Without considering the capacity constraints of the transmission system, the optimal pricing strategy problem of selling company i can be described as the problem of solving the maximum profit model of selling company i. On the premise of satisfying equations (7) and (8), the maximum value of equation (9) is obtained:

$$\sum_{i=1}^{M}P_i=D \tag{7}$$

$$P_{i\min} \le (1-k)P_i \le P_{i\max} \quad (8)$$

$$\max f_1 = \max(\xi_x P_i - C_i(P_i)) = \max\{\xi_x P_i - [0.5a_i(1-k)^2 P_i^2 + [b_i(1-k)+\rho k]P_i + c_i]\} \quad (9)$$

Where D is the predicted system load, equation (7) is the active power balance constraint of the system, and equation (8) is the power sales constraint of the power selling company. By solving the above model, we can determine the node real-time electricity price ξ_x and the electricity sold by each electricity selling company under unconstrained conditions.

3.3 The bidding model of electricity selling company with constraint

After the unconstrained scheduling between the power trading and dispatching center, the system parameters and the expected load level of each node are announced, and the power selling company i conducts power flow check on the results. The DC power flow check model is as follows:

$$\theta = B_0^{-1}P, \qquad P_{ij} = \frac{\theta_i - \theta_j}{x_{ij}} \quad (10)$$

Where, θ is the node voltage phase Angle vector; B_0 is the admittance matrix; P is the node net injection power vector; P_{ij} is active power flow of line; x_{ij} is line reactance; θ_i, θ_j are the voltage phase angles of the nodes at both ends of the line.

The characteristic of the tributary power flow method is to replace the AC power flow (active power and reactive power) of the power system with the equivalent DC current. The derivation of the above formula is shown below. Figure 2 shows the branch model.

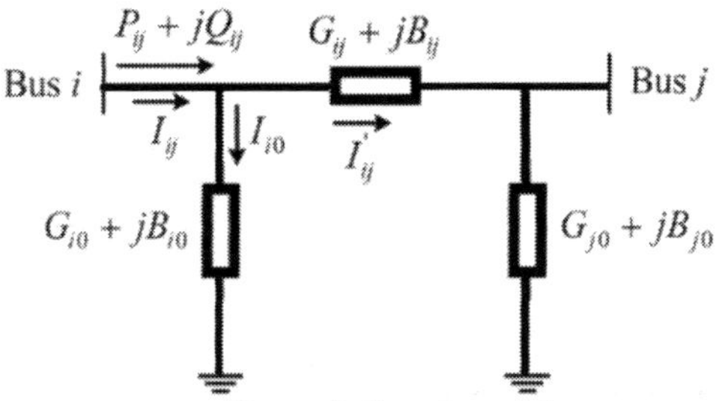

Figure 2. Branch model.

From the figure above:

$$P_{ij} + jQ_{ij} = \dot{U}_i \dot{I}_{ij}^* = \dot{U}_i(\dot{I}_{i0}^* + \dot{I}_{ij}'^*) \quad (11)$$

$$\begin{cases} \dot{I}_{i0} = \dot{U}_i(G_{i0} + jB_{i0}) \\ \dot{I}'_{ij} = (\dot{U}_i - \dot{U}_j)(G_{ij} + jB_{ij}) \end{cases} \quad (12)$$

Also known

$$\begin{cases} \dot{U}_i \dot{U}_i^* = U_i^2 \\ \dot{U}_i \dot{U}_j^* = U_i U_j \angle \theta_{ij} = U_i U_j(\cos\theta_{ij} + j\sin\theta_{ij}) \end{cases} \quad (13)$$

By combining formula (11), formula (12) and formula (13), we get:

$$\begin{cases} P_{ij} = U_i^2(G_{i0} + G_{ij}) - U_i U_j(B_{ij}\sin\theta_{ij} + G_{ij}\cos\theta_{ij}) \\ Q_{ij} = -U_i^2(B_{i0} + B_{ij}) + U_i U_j(B_{ij}\sin\theta_{ij} - G_{ij}\cos\theta_{ij}) \end{cases} \quad (14)$$

According to the need of DC power flow calculation, P_{ij} is simplified in three ways:

(1) Branch reactance is much larger than branch resistance, ignore branch resistance; At the same time, the conductance to the ground can also be ignored. So we have $G_{ii} = 0$, $G_{ij} = 0$;

(2) When calculated according to per unit value, the expected rated voltage of node voltage has little difference, so there is $|U_i| = |U_j| = 1$;

(3) The voltage phase Angle difference $(\theta_i - \theta_j)$ at both ends of the line is small, so there is $sin\theta_{ij} = \theta_i - \theta_j$, $cos\theta_{ij} = 1$.

The P_{ij} in the simplified formula (14) is:

$$P_{ij} = -B_{ij}\sin\theta_{ij} \quad (15)$$

The formula (15) is expressed in polar coordinates, thus:

$$P_i = \sum_{j=1}^{N}(B_{ij}(\theta_i - \theta_j)) = \sum_{j=1, j\ne i}^{N} B_{ij}\theta_i - \sum_{j=1, j\ne i}^{N} B_{ij}\theta_j \quad (16)$$

The above formula can be written in matrix form:

$$P = B\theta \quad (17)$$

Since $G_{ij} = 0$, then $jB_{ij} = 1/jx_{ij} = -j/x_{ij}$. Formula (15) can be rewritten as:

$$P_{ij} = -B_{ij}\sin\theta_{ij} = \frac{\theta_{ij}}{x_{ij}} = \frac{\theta_i - \theta_j}{x_{ij}} \quad (18)$$

The DC power flow checking model can be obtained.

If transmission congestion occurs, this pricing strategy needs to be adjusted. At this time, the real-time electricity price ξ_x of nodes in the system may be different, and the electricity selling cost function of the electricity selling company is:

$$W_2 = \sum_{i=1}^{M} \xi_i(P_i^0 + \Delta P_i) = \sum_{i=1}^{M} \lambda_i[\alpha_i + \beta_i(P_i^0 + \Delta P_i)] \cdot (P_i^0 + \Delta P_i) \quad (19)$$

In the formula, P_i^0 and ΔP_i are respectively the original electricity sold and the adjusted amount sold by the electricity selling company i.

After transmission congestion occurs, the maximum profit function of power selling company i is:

$$\max f_2 = \max(\xi_i(P_i^0 + \Delta P_i) - C_i(P_i^0 + \Delta P_i)) = \max\{\lambda_i[\alpha_i + \beta_i(P_i^0 + \Delta P_i)] \cdot (P_i^0 + \Delta P_i) - 0.5a_i(1-k)^2(P_i^0 + \Delta P_i)^2 + [b_i(1-k) + \rho k](P_i^0 + \Delta P_i) + c_i\} \quad (20)$$

3.4 Transmission congestion management model

The transmission line is the channel that sends the electric energy from the generator to the user, because the transmission line always has its transmission capacity limit, when the power flow of a transmission line reaches or exceeds its transmission capacity limit, the transmission congestion phenomenon occurs. The available transmission capacity (ATC) of a line reflects the difference between the line's capacity limit and its current power flow in the current operating state. Obviously, the phenomenon that the line is blocked is the case of ATC 0.

If unconstrained scheduling causes transmission congestion, then the power trading and dispatching center needs to carry out congestion management and require the power selling company to modify its bidding strategy to eliminate the congestion. In this paper, the transmission congestion management model is established with the aim of minimizing the congestion management cost, that is, the difference of the cost of electricity sales is minimized before and after the company modifies the bidding strategy. The problem can be described as follows:

$$\begin{aligned}\min f_3 &= \min(W_2 - W_1) \\ &= \min[\sum_{i=1}^{M}\xi_i(P_i^0+\Delta P_i)-\sum_{i=1}^{M}\xi_x P_i] \\ &= \min[\sum_{i=1}^{M}\lambda_i[\alpha_i+\beta_i(P_i^0+\Delta P_i)]\cdot(P_i^0+\Delta P_i)-\sum_{i=1}^{M}\xi_x P_i]\end{aligned} \quad (21)$$

$$\sum_{i=1}^{M}(P_i^0+\Delta P_i)=D, \quad 即\sum_{i=1}^{M}\Delta P_i=0 \quad (22)$$

$$P_{i\min} \le (1-k)P_i^0+\Delta P_i \le P_{i\max} \quad (23)$$

$$\theta = B_0^{-1}P, \quad P_{ij}=\frac{\theta_i-\theta_j}{x_{ij}} \quad (24)$$

$$P_{ij}' \le P_{ij\max} \quad (25)$$

Where, P_{ij}' and $P_{ij\max}$ are the active power flow of branch $i-j$ after congestion management and the transmission capacity limit of the line.

Equation (22) is the power balance constraint. Formula (23) is the constraint of electricity sales. Equation (24) is the power flow constraint. Equation (25) is the capacity constraint of transmission network.

The algorithm flow for the problem presented in this section is shown as follows:

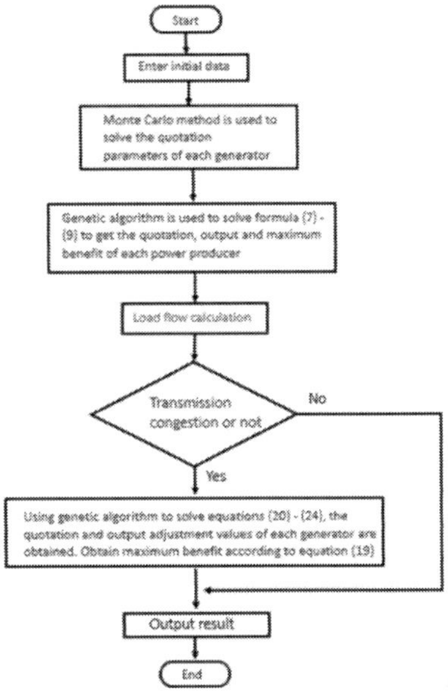

Figure 3. Algorithm flow chart.

4. Analysis of examples

The standard IEEE 14-node system is taken as an example to illustrate the method proposed in this paper. Figure 4 shows the wiring diagram of the IEEE 14-node system, and the system parameters are shown in reference.

Assume that the total demand load announced by the market $D=369MW$. Nodes 1, 2, 3, 6 and 8 are power selling companies established by five traditional thermal power producers. Assuming that each power producer has only one generator set, the proportion of renewable energy quota k is 15%, and the parameters of power selling companies are shown in Table 1. Taking company 2 as the research object, it is assumed that the quotation parameters of other competitors follow the normal distribution form, and the quotation parameters of competitors are independent of each other, and the values of expected value and variance are $\mu_i=1.2,\ \sigma_i=0.15,\ (i=1,3,4,5)$.

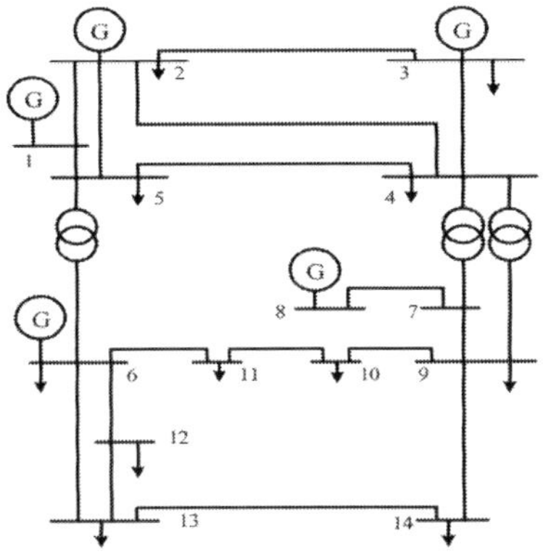

Figure 4. IEEE 14 node system wiring diagram.

Table 1. Electricity selling company parameters

Electricity sales company i	a_i $\$/(MW^2h)$	b_i $\$/(MWh)$	c_i $\$/h$	Minimum electricity sales $/MW$	Ceiling of electricity sales $/MW$	Node number
1	0.0750	4.00	0	15	100	1
2	0.1375	3.00	0	15	90	2
3	0.0525	5.25	0	20	100	3
4	0.0875	3.00	0	20	100	6
5	0.0750	5.00	0	20	120	8

Table 2. The amount of electricity sold by each electricity selling company without constraint

Electricity sales company i	1	2	3	4	5
power sale quantity $P_i\,/\,MW$	88.011	53.021	76.188	72.845	78.935

When MC method is used to randomly sample the quotation parameters of competitors, the given number of random simulations is T =3000. In each simulation, the bidding parameters of the competitor are fixed. The population size of GA is 50, the maximum allowable genetic algebra is 50, and the crossover and mutation probabilities are 0.9 and 0.005, respectively. Here, it is required to solve the optimal quotation parameter λ_2 of the selling power company 2, which is a one-place search problem of a single optimization parameter, because the actual quotation curve of the selling power company is higher than the marginal cost curve, and the search interval given λ_2 is [1,2.98].

According to the calculation process shown in Figure 4.3, the optimal quotation parameters of power selling company 2 are λ_2 * =1.086, on-grid power P_2 * =53.021MW, the optimal quotation $\xi_2 * = 10.933\$/MW$, and the maximum profit is 185.539$ without considering the transmission capacity constraints. The electricity sold by the corresponding electricity selling companies is shown in Table 2.

Based on the above results, the power flow is calculated in the network. It is found that branch 4 and Branch 12 are congested. Table 3 lists the specific data.

Table 3. The branch transmission is blocked

Branch number	Bus end number	Branch power flow (per-unit value)	Upper limit of branch power flow(per-unit value)
4	1-5	0.594	0.4
12	6-12	0.378	0.2375

After transmission congestion occurs, power selling companies adjust their own bidding strategy and electricity sales, then solve the transmission congestion model, and get the output adjustment value of each power selling company after management. For electricity selling company 2, the adjusted value of electricity sold $\Delta P_2 * = 21.609MW$, then the offer $\xi *' = 13.264\$/MW$, the maximum profit is 355.046$.

It can be seen that the bidding strategy formulated by the power selling company 2 after the transmission congestion management greatly improves the profit compared with the unconstrained situation. The corresponding adjustment of the electricity sales of each electricity selling company is shown in Table 4, positive value means that the electricity sales increase, negative value means that the electricity sales are reduced.

Table 4. The adjusted value of electricity sold by each electricity selling company

Electricity sales company i	1	2	3	4	5
Electricity sales adjustment value P_i / MW	-45.531	21.609	-15.068	-1.965	40.965

In the case of transmission congestion management, we analyze the change of bidding strategy of electricity selling companies when the impact of renewable energy quota and green certificate trading is not considered. At this time, excluding the transaction cost of green certificate ($k=0$), the cost of electricity sold by the electricity selling company will change, and the quotation will also change, which will cause the change of electricity sold. For power selling company 2, the optimal offer parameter is $\lambda_2{*}'=1.194$, power selling $P_2{*}=78.70MW$, the optimal offer is $\xi_2{*}' =16.502\$/MW$, and the maximum profit is 636.791$. The electricity sold by the corresponding electricity selling companies is shown in Table 5.

Table 5. Sales volume of electricity companies before and after considering green certificate costs under transmission congestion management

Electricity sales company i	1	2	3	4	5
Electricity sold including the cost of green card P_i / MW	42.48	74.63	61.12	70.88	119.99
Electricity sold without considering the cost of green certificates P_i / MW	39.11	78.70	63.94	67.35	119.99

Through comparison, it can be clearly and directly found that when executing the optimal quotation strategy of power selling company 2, the output of each power selling company does not change significantly without considering the impact of renewable energy quota and green certificate transaction, which mainly affects the quotation and profit of power selling company 2, both of which have significantly increased.

5. Conclusions

Under the influence of the new policy of renewable energy quota and green certificate trading, the optimal bidding model of power selling companies established by traditional thermal power producers is established, taking into account transmission congestion. Through research and analysis, it can be seen that when the electricity selling company includes the green certificate transaction cost into the electricity selling cost, it can meet the policy demand and make the model more accurate. When formulating the optimal bidding strategy, quantifying the newly issued green certificate trading policy, including the cost of selling electricity, using probability function to describe the bidding behavior of competitors, and considering the possible network congestion before quoting, not only can improve the success rate of a transaction, but more importantly, can obtain higher profits, which proves the practicability and reliability of this method.

Acknowledgments

This work was supported by the Science and Technology Project of SGCC, Research on Electricity market Clearing Model and Trading System considering source load Uncertainty (5217L022001G).

References

1. WANG T, GONG Y, JIANG C. A review on promoting share of renewable energy by green-trading mechanisms in power system[J]. Renewable and Sustainable Energy Reviews,2014, 40:923-929.
2. An Xuena, Zhang Shaohua, Li Xue. Equilibrium analysis of oligopolistic electricity markets considering tradable green certificates[J]. Automation of Electric Power Systems, 2017, 41(9): 84-89. DOI:10.7500/AEPS20160516008.
3. Zhang Xiaohui, Yan Pengda, Zhong Jiaqing, et al. Low carbon economic power planning under the renewable energy incentive system [J]. Power Grid Technology,2015,39(3):655-662.
4. Zhang Boming, Chen Shousun. Advanced Power Network Analysis [M]. Tsinghua University Press, 2007.
5. Tang Zhenfei, Tang Guoqing, Yu Erkeng. Management of transmission congestion in the electricity market[J]. Power System Automation, 2001, 25(23): 13-16. DOI:10.3321/j.issn:1000-1026.2001.23.004.

Analysis of common reservoir damage problems in coalbed methane wells and study on acid plugging removal agent system

Jin Liao, Xianjun Li, Jingchen Han, Zhou Yang, Shuling Zhang

Gepetto Petroleum Technology Group Co., Ltd., Chengdu 610500, China

Abstract. Reservoir damage is an important factor limiting the productivity of coalbed methane wells. In this paper, the coal seam in a coalbed methane producing area in the southeast edge of Ordos Basin is taken as the research object, and the types of reservoir damage in this block are analyzed through the composition test of coal samples and the comparison of production conditions, and the targeted solutions are given. The results show that the main types of reservoir damage in this block are solid plugging and water locking damage. The dissolution rate of acidic plugging removal working fluid with hydrochloric acid, hydrofluoric acid, acetic acid and hydrogen peroxide as main components for pulverized coal in this block reached 13.73%. After acidification, the permeability increased by 1.8~6.9 times, and the porosity increased by 0.35%~2.87%, indicating that the plugging removal working fluid system can effectively dredge the liquid seepage channel, reduce the seepage resistance and achieve the plugging removal effect.

Keywords: Coalbed methane; reservoir damage; acidic plugging remover; experimental evaluation.

1. Introduction

China is a big coal producer, and as an associated mineral of coal, the resources of coalbed methane are also very considerable[1]. According to statistics, China's coalbed methane resources with a buried depth of less than 2000m are about 3.7 billion cubic meters, equivalent to 35 billion tons of standard oil, which has great development potential. However, although the country attaches great importance to the development of coalbed methane industry and gives policy support, three consecutive five-year planning targets of coalbed methane industry have not been achieved, among which the surface coalbed methane output in 2020 is about 6 billion cubic meters, far below the 10 billion cubic meters required in the Thirteenth Five-Year Plan for the Development and Utilization of Coalbed Methane (Coal Mine Gas), and the target completion rate is only about 60%.

There are many reasons for this situation, among which reservoir damage is one of the important limiting factors that restrict the productivity of coalbed methane wells. Affected by geology and production, the fluid seepage velocity in coal reservoir is low, which leads to the reservoir damage problems such as water lock and solid phase blockage in the production process of coalbed methane wells, further limiting the production, resulting in low production or even shutdown of gas wells. Taking a block in the southeastern margin of Erduo Basin as an example, the average daily gas production of more than 50 coalbed methane wells in this block is less than 100 cubic meters due to reservoir damage. If the plugging can be successfully removed, the daily gas production can exceed 20,000 cubic meters, with remarkable benefits.

It is an important way to jointly tackle key problems in large areas and develop targeted technologies according to local conditions. In this paper, taking a coalbed methane block in the southeastern margin of Ordos Basin as an example, the causes of reservoir damage are analyzed, and a targeted plugging removal agent system is developed based on the coal samples in this block.

2. Cause Analysis of Reservoir Damage in Coalbed Methane Wells

2.1 Coal Sample Analysis

Taking the 5# coal seam in a coalbed methane block in the southeast edge of Ordos Basin as the research object, the coal samples were analyzed, which were taken from a coal producing area about 500 meters away from a coalbed methane production well (also from 5# coal seam). The mineral composition and characteristics of porosity and permeability system of the samples were analyzed, and the results were recorded as follows.

(1) Microscopic analysis

The test coal samples were crushed and made into optical sheets[2], and the vitrinite reflectance was measured by LV100PO polarized microscope. The test results are shown in Figure 1.

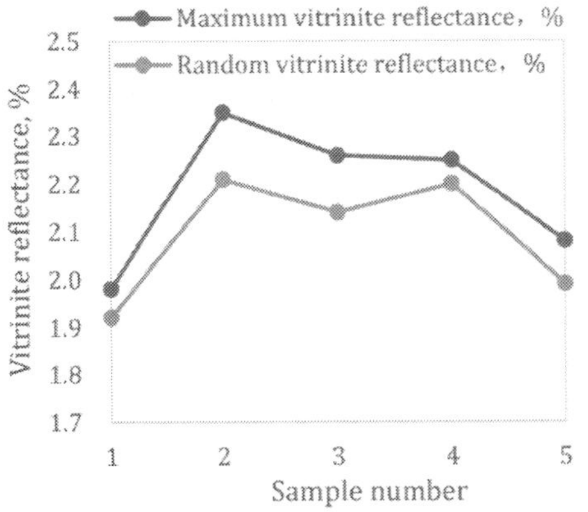

Figure 1: Test Results of Vitrinite Reflectance of Coal Samples

According to the experimental results, the maximum vitrinite reflectance of coal samples is between 1.98% and 2.35%. According to the relevant industry standards, all coal samples belong to middle-rank coal VII[3]. Generally speaking, low coalification will lead to low organic matter content, and it is difficult for methane enrichment to reach the level of industrial development; Excessive coalification will lead to poor pore connectivity, which is not conducive to methane gas seepage. The experimental coal sample belongs to middle-high rank coal with moderate coalification degree, which is a favorable condition for coalbed methane development.

(2) Mineral composition test

Based on X-ray diffraction whole rock analysis technology, the mineral composition of coal samples in this block was tested, and the test results obtained according to diffraction patterns are shown in Table 1.

Table 1: Test Results of Mineral Components.

Serial number	Mineral composition, %								
	clay mineral	barite	quartz	potassium feldspar	plagioclase	calcite	dolomite	siderite	pyrite
6	79.6	—	—	—	—	9.3	11.1	—	—
7	64.7	—	—	—	—	14.2	17.6	—	3.5
8	77.3	—	3.1	—	—	6.9	12.7	—	—

According to Table 1, the mineral components of coal samples in this block are mainly clay minerals, accounting for about 60%~80%, in addition to some carbonate minerals such as calcite and dolomite. According to the experience of acidizing operation, hydrofluoric acid and hydrochloric acid, which are commonly used in acidizing, can effectively dissolve the above minerals. Therefore, when the reservoir damage caused by mineral migration occurs in the coal reservoir, it is theoretically feasible to remove the plugging by acidification.

(3) Determination of characteristic parameters of porosity and permeability

The porosity and permeability characteristics of coal samples are closely related to the storage and seepage of coalbed methane, and directly related to the production of coalbed methane. Poor porosity and permeability parameters will increase the risk of reservoir damage. The coal sample is made into a cylinder with a diameter of 25.0mm and a height of 40.0~50.0mm, and the porosity and permeability parameters of the coal sample in the block are measured by using a gas porosity and permeability tester with nitrogen as the medium. The test results are shown in the following table.

Table 2: Measurement Results of Porosity and Permeability Parameters of Coal Samples

Serial number	length, mm	diameter, mm	porosity, %	permeability, mD
H1	50.0	25.0	4.58	0.023
H2	48.2	25.0	11.52	0.031
H3	47.9	25.0	7.38	0.095
H4	42.6	25.0	5.33	0.117
H5	50.0	25.0	4.91	0.143
H6	44.7	25.0	5.84	0.027
H7	48.9	25.0	6.63	0.058
H8	50.0	25.0	6.85	0.083

According to the test results, the porosity of coal samples in this block is between 4% and 12%, which varies greatly among different coal samples, but the porosity is mostly around 5%. The permeability mainly fluctuates between 0.02 and 0.15 mD, and the permeability of most coal samples is lower than 0.1 mD. On the whole, the coal samples in this block have typical characteristics of "low porosity and low permeability".

2.2 Cause Analysis of Reservoir Damage in Coalbed Methane Wells

According to the test results of coal samples, the reservoir in the study block is medium-high rank coal with moderate coalification degree, and the abundance of coalbed methane reserves meets the requirements of industrial development. However, the coal seam in this block has typical characteristics of "low porosity and low permeability" and clay minerals are developed, so the reasons for reservoir damage of coalbed methane wells in this block are summarized as follows:

(1) Water lock damage caused by polymer adsorption and retention, capillary force (resistance) and so on. A large number of polymer working fluids, such as guanidine gum and polyacrylamide, are used in fracturing coalbed methane wells. Due to the low formation pressure and insufficient gel breaking in coalbed methane wells, these polymers cannot be completely discharged back to the ground, resulting in retention and water lock damage. In addition, according to the capillary force formula, the smaller the pore throat radius and the more hydrophilic the coal petrography are, the greater the seepage resistance of liquid will be. In general, coal petrography are more hydrophilic, and the coal reservoir in the research block has the typical characteristics of "low porosity, low permeability and low pressure", which leads to the large seepage resistance of liquid in the production process, and the liquid is easy to block at the narrow throat, resulting in water lock damage.

(2) Water sensitivity damage caused by incompatibility between foreign fluids and reservoir rocks. During the drilling and reservoir reconstruction of coalbed methane wells, a large amount of working fluid will enter the reservoir. Generally speaking, clay stabilizer and iron ion stabilizer will be added to these working fluids to prevent reservoir damage. However, it is difficult for foreign fluids not to damage the reservoir completely, especially for layer with high clay mineral content, water-sensitive damage is almost inevitable. After water-sensitive damage, clay minerals absorb water, swell, disperse and migrate, which leads to narrowing or even blocking the throat and aggravates the damage of water lock.

(3) Solid plugging caused by solid particle migration. In the production process of coalbed methane wells, coal powder, clay minerals and sediments will migrate with liquid seepage due to the influence of hydraulic scouring and stress change. These solid particles are easy to gather in the narrow throat, and then block the migration channel. The produced water of coalbed methane wells in the research block is mostly black water or slightly black water, and solid particles such as coal powder can be seen after standing, so it can be judged that the reservoir damage caused by solid plugging is common in this block.

3. Feasibility Analysis of Acid Plugging Removal in Coal Reservoir

According to experimental research and production experience, water lock damage and solid plugging are important factors that limit the productivity of coalbed methane wells in the research block, and it is difficult to solve the above problems simultaneously by traditional physical plugging removal methods. The purpose of plugging removal is to dredge the fluid seepage channel and reduce the fluid seepage resistance. According to the capillary force formula, the ways to reduce seepage resistance mainly include reducing surface tension, increasing contact angle and increasing throat radius. Therefore, theoretically speaking, using acidic working fluid and selecting appropriate additives can realize the unblocking of coal seams in this block.

(1) The coal petrography minerals in the study area are mainly clay minerals and carbonate minerals, with a small amount of quartz and pyrite. According to scholars' research results and acidizing experience, these minerals can be dissolved by hydrochloric acid, acetic acid and hydrofluoric acid, and the carboxyl group (-COOH) in acetic acid can form a complex with iron ions through coordination to prevent secondary precipitation. The reaction equation between acidic working fluid and coal seam minerals is as follows(Where M is a metal ion) [5-6]:

$$\mathrm{MCO_3}(Carbonate\ minerals) + \mathrm{HCl} \rightarrow \mathrm{MCl} + \mathrm{H_2O} + \mathrm{CO_2}\uparrow \quad (1)$$

$$\mathrm{MCO_3} + \mathrm{HA_c} \rightarrow \mathrm{MA_c} + \mathrm{H_2O} + \mathrm{CO_2}\uparrow \quad (2)$$

$$\mathrm{Fe_2O_3} + \mathrm{HCl} \rightarrow \mathrm{FeCl_3} + \mathrm{H_2O} \quad (3)$$

$$\mathrm{Al_2Si_2O_5(OH)_4}(kaolinite) + \mathrm{HF} \rightarrow \mathrm{AlF_3} + \mathrm{H_2SiF_6} + \mathrm{H_2O} \quad (4)$$

$$\mathrm{AlO_3} \cdot 4\mathrm{SiO_2} \cdot \mathrm{H_2O}(montmorillonite) + \mathrm{HF} \rightarrow \mathrm{H_3AlF_6} + \mathrm{H_2SiF_6} + \mathrm{H_2O} \quad (5)$$

$$\mathrm{SiO_2} + \mathrm{HF} \rightarrow \mathrm{SiF_4} + \mathrm{H_2O} \quad (6)$$

$$\mathrm{SiF_4} + \mathrm{HF} \rightarrow \mathrm{H_2SiF_6} \quad (7)$$

(2) Oxidant can dissolve part of coal matrix through oxidation, and release heat, carbon dioxide and organic acid at the same time, so as to dredge seepage channels and improve reservoir porosity and permeability. Jing Zhenhua and others compared and analyzed the oxidation effects of hydrogen peroxide, potassium permanganate, sodium hypochlorite and potassium persulfate on coal powder when studying the mechanism of oxidation and infiltration enhancement of coal reservoir. The results showed that all four oxidants could react with coal powder, and potassium permanganate and sodium hypochlorite had the best dissolution effect on coal powder. However, the reaction product of potassium permanganate contains manganese dioxide which is insoluble in acid, and sodium hypochlorite will generate toxic chlorine gas in acidic solution, which are not suitable for adding into acidic plugging remover. The oxidation of hydrogen peroxide will be more intense under acidic conditions, and the reaction product is mainly water, which has high environmental protection value. At the same time, hydrogen peroxide can destroy the chemical bonds on the coal surface and promote the desorption of methane molecules, which is a good choice for oxidizing coal matrix.

(3) Fluorocarbon surfactants can effectively improve the interface parameters. Fluorocarbon surfactant is one of the

commonly used surfactants in oil and gas fields at present. Compared with conventional surfactants, it has better effect of reducing surface tension and improving wettability, and can effectively reduce liquid phase seepage resistance. In addition, fluorocarbon surfactants have good acid resistance and is a good choice for acidic working fluids when the cost allows.

4. Experimental Evaluation of Working Fluid for Acid Plugging Removal of Coal Petrography

With hydrochloric acid, hydrofluoric acid, acetic acid and hydrogen peroxide as the main components, a four-factor and three-level orthogonal experiment was designed for single-dose compounding. After the orthogonal experiment was completed, the effect was analyzed with the dissolution rate of pulverized coal as the main evaluation index, and the main formula of acid solution was obtained. Fluorocarbon surfactant A was selected as the working fluid additive, and finally the acidic plugging removal system JD1-1 was obtained. Taking the coal sample in the research block as the experimental object, the indoor evaluation experiment of acid plugging removal system JD1-1 was carried out, and the results are as follows.

(1) Pulverized coal dissolution experiment

The working fluid JD1-1 is mixed with dry pulverized coal in a plastic beaker according to the ratio of 30ml: 3g, and fully stirred with an inert plastic rod to make the acid liquid fully contact with pulverized coal. After 12 hours, the acid solution mixed with pulverized coal in the beaker was filtered with fine filter paper to obtain wet pulverized coal. Drying the wet pulverized coal to obtain the weight after acid solution dissolution and calculate the dissolution rate. After repeated experiments for three times, the dissolution rate of acidic plugging remover JD1-1 for pulverized coal in this block is 13.05%~14.30%, with an average of 13.73%, which shows that the dissolution effect is good.

Table 3: Test Results of Dissolution Rate

Serial number	Quality before treatment, g	Quality after treatment, g	Lost quality, g	Dissolution rate, %
1	30.00	26.09	3.92	13.05
2	30.00	25.85	4.15	13.84
3	30.00	25.71	4.29	14.30

(2) Evaluation experiment of porosity and permeability parameters before and after acidification

The core was soaked in acidic working fluid JD1-1 for 12 hours, dried at 60°C, and the porosity and permeability parameters after acidification were measured by gas porosity and permeability tester. The experimental results show that after acidification, the core porosity increases by 0.35%~2.87%, the permeability increases by 1.8~6.9 times, and the porosity and permeability parameters are significantly improved.

Table 4: Test Results of Dissolution Rate

Serial number	Length,mm	Diameter,mm	Porosity before acidification,%	Permeability before acidification, mD	Porosity after acidification,%	Permeability after acidification, mD
1	25	43.7	6.14	0.045	9.01	0.356
2	25	49.3	8.53	0.091	8.88	0.255
3	25	45.9	7.42	0.074	8.49	0.221

5. Conclusion and Recommendations

(1) The coal petrography minerals in the study area are mainly clay minerals and carbonate minerals, with a small amount of quartz and pyrite.Solid plugging and water lock damage are the main reasons for reservoir damage of coalbed methane wells in this block.

(2) The acid plugging removal system JD1-1 was developed according to the damage causes of the research block. The test results show that the working fluid can effectively dissolve pulverized coal and improve the porosity and permeability parameters of coal core.

(3) The plugging removal effect of acid plugging removal system JD1-1 is only aimed at the coal samples in the study block, and has not been verified by field tests. It is

suggested to develop a more universal plugging removal system for more coal samples in the future, and to carry out field tests to verify the plugging removal effect.

References

1. Zhao Weibo, Liu Honglin, Wang Huaichang, et al. Controlling Effect of Coal Relative Pore Structure-Taking Yulin 8# Coal in Ordos Basin as An Example [J/OL]. Coal Science and Technology: 1-17.
2. Wang Yunfei. Study on Harnessing Low-yield and Low-efficiency CBM Wells in Block H on the Eastern Edge of Hubei [D]. Xi 'an Shiyou University, 2022.
3. Liu Jinrong. Adsorption and diffusion law and control mechanism of low rank coalbed methane in southeastern Yunnan [D]. China University of Mining and Technology (Beijing), 2018.
4. Liu Yucong. Mechanism and Laboratory Experimental Study of Surfactants in Coalbed Methane Reservoir [D]. Southwest Petroleum University, 2019.
5. Xiao Zhiguo, Hao Mei. Research status and progress of acidification and permeability enhancement technology in coal seam [J]. Coal Mine Safety, 2023, 54(10):1-7.
6. Hu Qianting, Li Xiaoxu, Chen Qiang, et al. Experimental study on acid fracturing fluid to prevent water lock damage in low permeability coal seam [J]. Acta Sinica, 2022,47(12):4466-4481.
7. Jing Zhenhua, Pan Songqi, Wang Xiaoming, et al. Mechanism and application of oxidation and infiltration enhancement in coal reservoirs [J]. Acta Sinica Coal, 2022,47 (11): 3975-3989.

Analysis and research on the chemical composition of ancient Chinese glass

Ming Fang [1,*], Jiaqi Huang [2]

[1] School of Computer Science, Guangdong University of Technology, Guangzhou, 510006, China
[2] School of Foreign Language, Guangdong University of Technology, Guangzhou, 510006, China

Abstract. The purpose of this study is to investigate the classification rules for lead-barium and high-potassium glass. First, we noticed that the glass types differed considerably for the five components. After analyzing the data and establishing the independent T-test model, We developed the Logistic regression classification model using these components as independent factors and the glass types as dependent variables, and we derived the classification function for the glass types. Second, high-potassium glass, was divided into two subclasses based on calcium and aluminum content using the K-means++ model and associated literature on the chemical composition of ancient Chinese glass [1][2]. These were grass-ash high-potassium glass and nitrate high-potassium glass, as well as two types of lead-barium glass: lead-sodium glass and lead-calcium glass. Finally, the cluster center coordinate values were perturbed by -10% and 10%, verifying the classification of each sample under different perturbations, and the resulting subclass divisions remained unchanged, confirming the low sensitivity of the classification results.

Keywords: logistic regression, K-means++ model, sensitivity analysis

1. Restatement of the issue

We statistically examined the classification rule of high-potassium glass and lead-barium glass for the obtained data, and then used the clustering technique to determine the proper chemical composition for each category to divide them into subclasses. Second, we provided clear and specific classification methods as well as reasonable classification findings based on clustering results and mathematical statistical analysis, and then assessed the rationality of the classification results in light of objective laws and the real circumstances [3][4]. Finally, we introduced noise to see how sensitive the classification findings are.

2. Analysis of the issue

To analyze the classification rules for analyzing high-potassium glass and lead-barium glass, we first combined the data and information we gathered, and then we performed the T-test of independence for each type of chemical composition, yielding five important factors such as SiO_2 and K_2O [5][6]. These separated components were employed as independent factors and glass type as dependent variables to create a logistic regression classification model based on the statistical laws established. We selected the proper chemical composition for each category to divide it into subclasses and provided the precise division method and division results. And then, we investigated high-potassium glass and lead-barium glass and performed independent T-tests on their chemical compositions. As a consequence, we obtained the importance of alumina and calcium oxide in high-potassium glass, as well as the importance of atrium oxide and calcium oxide in lead-barium glass, and we applied the K-means++ model for clustering analysis. Using K-means++ model clustering analysis, we discovered that the subclass of high-potassium glass can be separated into two classes. Based on a survey of related literature[7][8], the subclass of lead-barium glass can be classified into two groups, each of which is dominated by two elements: lead and lead calcium. At the same time, we used the literature and study reports from linked researchers to demonstrate the validity of the classification results. To test the sensitivity of our model to the classification findings, we introduced perturbations to the cluster center coordinates.

3. Modeling and issue solving

3.1 Exercise independence on extensive data to derive crucial components

In the Table 1, the F value represents the ratio of the between-group mean square, while the value denotes the level of significance. Because F > 1 for all tabular entries

* Corresponding author: Ming Fang

in the table, there was a difference between the groups, and the larger the F value, the larger the difference. Furthermore, $\rho<0.05$ shows that the analysis was statistically significant for all items in the table. As a result of the data in Table 1 and the foregoing analysis, five essential factors were obtained: SiO_2, K_2O, PbO, BaO, and weathering or not.

Table 1. Partial ANOVA results

Chemical composition	SiO_2	K_2O	PbO	BaO
F	50.332	29.063	2.842	2.160
ρ	0.000**	0.000**	0.011*	0.016*

3.2 The combined data's independence tests revealed significant components.

The detailed procedure for logistic regression modeling is as follows:

These significant variables with differences were employed as independent variables x to obtain

$$\theta_0 + \theta_1 x_1 + \cdots \ldots + \theta_n x_n = \theta^T x \tag{1}$$

We Substituted the Sigmoid function,

$$y = \frac{1}{1 + e^{-(\theta^T x + b)}} \tag{2}$$

Then we constructed the loss function,

$$J(\theta) = -\frac{1}{n}\left(\sum_{i=1}^{n}(y_i \ln p(x_i) + (1 - y_i)\ln(1 - p(x_i)))\right) \tag{3}$$

To update the parameters, the gradient descent approach was employed, and the final classification function was obtained as:

$$y = \frac{1}{1 + e^{-(4.768+0.185x_1+0.433x_2-2.52x_3-0.23x_4+9.237x_5)}} \tag{4}$$

Where to find SiO_2, K_2O, BaO, PbO, and weathering or not.

3.3 K-means++ Clustering

3.3.1 Difference value analysis of high-potassium glass

We discovered that the chemical composition content of calcium and aluminum in high-potassium glass that used grass ash as a combustion aid and nitrate as a combustion aid differed significantly after reviewing relevant information [9], so we first performed a difference test on the calcium and aluminum elements of high-potassium glass, the results of which were shown in Table 2.

Table 2. Difference test results of high-potassium glass

Chemical composition	F	ρ
Al_2O_3	32.426	0.000***
CaO	31.373	0.000***

Since the F-values of aluminum oxide and calcium oxide are much greater than 1, it means that there is a considerable difference in the clustering of calcium and aluminum components.

3.3.2 Analysis of the differential value of lead-barium glass

We analyzed the relevant information for lead-barium glass and discovered that in the ancient glass [10], there were two types of lead-sodium glass and lead-calcium glass, so we evaluated the difference between the sodium element and the calcium element, and the findings are displayed in Table 3 as follows:

Table 3. Difference test results of lead and barium glass

chemical composition	F	ρ
Na_2O	65.026	0.000***
CaO	2.048	0.047*

Similarly, the F-values of sodium oxide and calcium oxide are all greater than 1, indicating that there is a significant difference in the clustering of calcium and sodium components.

3.3.3 Constructing the K-means++ model

Using the K-means++ model, we clustered high-potassium glass and lead-barium glass, respectively. Using high-potassium glass as an example, the specific approach is as follows:

Step 1.

We generated the first dataset with calcium elements as x-axis coordinates and sodium elements as y-axis coordinates.

Step 2.

We chose a random sample point from the data set to serve as the initial clustering center (c_1).

Step 3.

The shortest distance between each sample and the currently existing clustering center, represented by D(x), was calculated first, followed by the likelihood that each sample were chosen as the next clustering center:

$$\frac{D(x)^2}{\sum_{x \in XD} D(x)^2} \tag{5}$$

Finally, the next clustering center was chosen based on the roulette wheel.
Step 4.
We repeated step 2 until a total of K clustering centers were selected.
Step 5.
We calculated the distance between each sample x_i in the dataset and assigned it to the class corresponding to the clustering center with the shortest distance.
Step 6.
We recalculated its clustering center for each class.

$$c_i = \frac{1}{|c_i|}\sum_{x \in c_i} x \tag{6}$$

Step 7.
We repeated Steps 5 and 6 until the position of the clustering center no longer changed.
The final clustering results of high potassium glass generated by the seaborn module are shown in Figure 1.

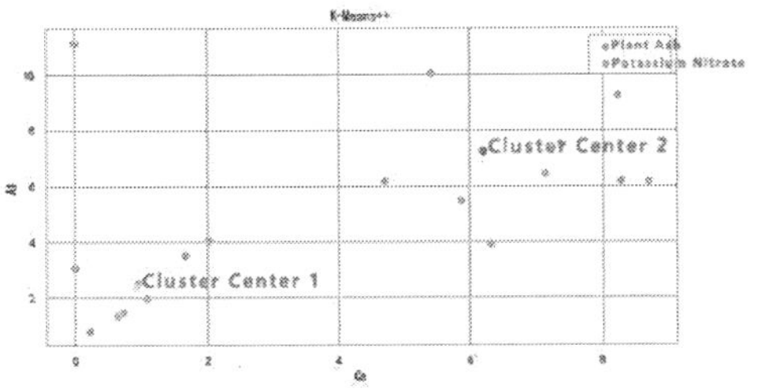

Figure 1. Clustering results of high-potassium glass

Figure 1 shows two clustering centers: the high-potassium glass clustering center of the grass ash class and the high-potassium glass clustering center of the nitrate class. The orange point represents the nitrate class's high-potassium glass, which is concentrated in the upper right of the coordinate axis, and the blue point represents the grass ash class's high-potassium glass, which is concentrated in the lower left of the coordinate axis. In the image, the two clusters are far apart, indicating that the clustering effect is efficient, and it makes sense to divide the subclass of high potash glass into the high potassium glass center of the grass ash class and the high potash glass center of the nitrate class.
Similarly, the clustering results of lead-barium glass generated by the seaborn module are shown in Figure 2.

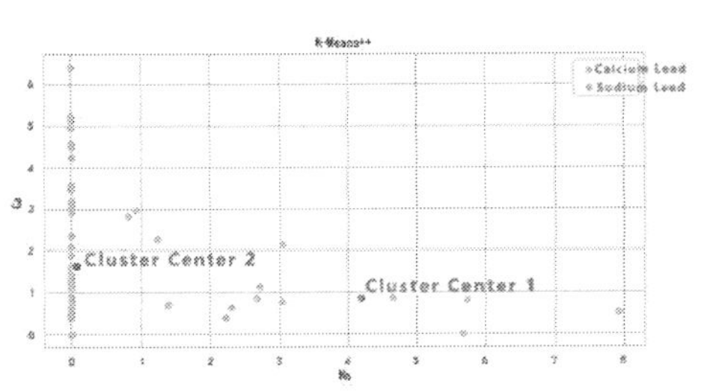

Figure 2. Clustering results of lead-barium glass

Figure 2 shows that there are two clustering centers: the lead-calcium glass clustering center and the lead-sodium high-potassium glass clustering center. The orange points on the left side of the axis depict the lead-calcium-lead-barium glass. The blue points on the right side of the axis represent the lead-sodium-lead-barium glass. The two clusters are widely apart in the image, indicating that the clustering effect is stronger, and it is meaningful to separate the lead-barium glass subclasses into lead-calcium lead-barium glass and lead-sodium lead-barium glass centers.

3.4 Sensitivity analysis

We reviewed the sensitivity of the classification results in order to verify the reliability and stability of our classification results. Our method detects the sensitivity of classification results by modifying the coordinate values of the clustering centers and applying a -10% -10% perturbation in this way. Figures 3 and 4 illustrate the sensitivity detection findings for the high-potassium glass:

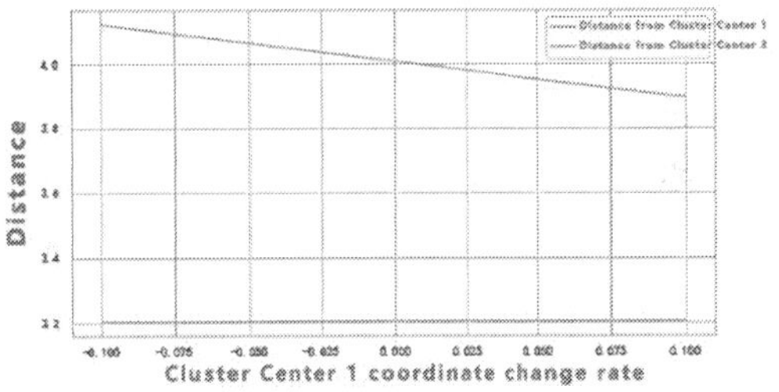

Figure 3. Sensitivity analysis of high-potassium glass clustering center 1

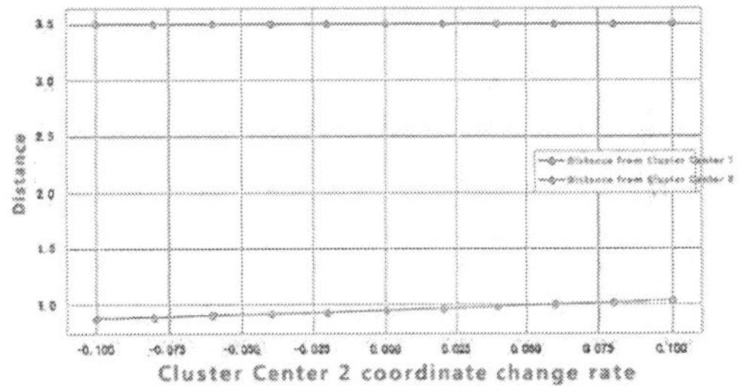

Figure 4. Sensitivity analysis of high-potassium glass clustering center 2

When the coordinates of high-potassium glass clustering center 1 are transformed, as shown in Figures 3 and 4, the distance between the high-potassium glass samples classified as nitrate class and high-potassium glass clustering center 1 gets closer and closer, but the clustering results remain unchanged, indicating that the results of the classification of nitrate class high-potassium glass do not change with the perturbation. Similarly, when the coordinates of the clustering center 2 of high-potassium glass are transformed, the grass-ash samples of high-potassium glass become increasingly distant from the clustering center 2, but the clustering results do not change, indicating that the results of the division of the grass-ash class of high-potassium glass do not change with the perturbation, proving that the subclass division model of high-potassium glass has low sensitivity, strong stability, and a good effect.

Similarly, the high potassium glass sensitivity detection results generated by the seaborn module are shown in Figures 5 and 6.

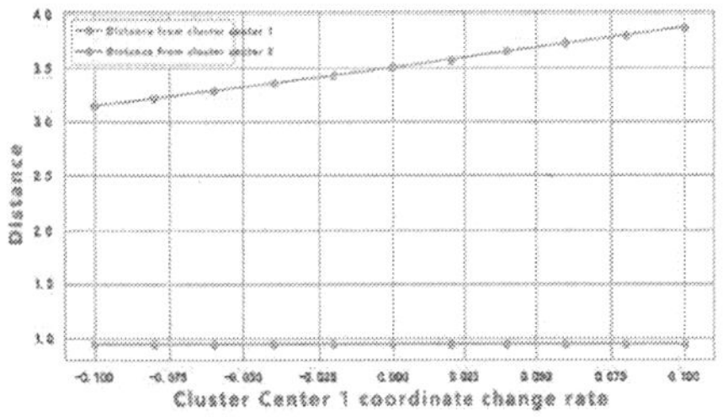

Figure 5. Sensitivity analysis of lead barium glass clustering center 1

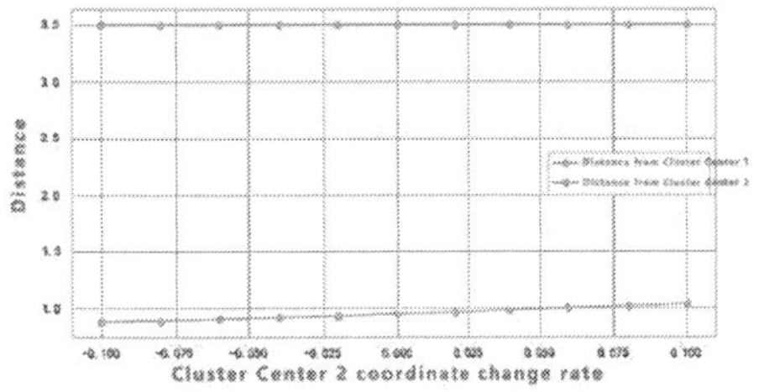

Figure 6. Sensitivity analysis of lead barium glass clustering center 2

When the coordinates of the clustering center of lead-barium glass are altered, as shown in Figures 5 and 6, the samples of lead-barium glass classified as lead-calcium lead-barium glass are farther apart. When the coordinates of lead-barium glass's clustering center 2 are transformed, the samples of lead-barium glass classified as lead-sodium lead-barium glass become increasingly distant from the clustering center 2 of lead-barium glass, but the clustering results do not change, indicating that the results of the division of lead-sodium lead-barium glass do not change with the perturbation. As a result, the low sensitivity of the subclassification model for lead-barium glass suggests the model's stability and efficacy.

4. Conclusion

In order to acquire the five factors of silica, potassium oxide, barium oxide, lead oxide, and weathering or not that exhibit significant differences in the classification law of high-potassium glass and lead-barium glass, we first utilized the independent T-test model. Second, to characterize the classification law of high potassium and lead-barium glass, a logistic regression model was developed. Then, using the classification function of glass types and the related ancient Chinese glass literature, we established the Kmeans++ model to classify high-potassium and lead-barium glass artifacts into two subclasses: high-potassium glass artifacts into two subclasses of grass-ash type high-potassium glass and nitrate-type high-potassium glass, and lead-barium glass artifacts into two subclasses of lead-sodium glass and lead-calcium glass. Finally, sensitivity analysis can be used to validate the model's validity and the classification's reasonableness. Further investigation into the challenges on this subject still requires effort.

References

1. Zeng, W.S. (2022). Ancient Chinese and Western glassware and their cultural differences as seen in the excavated glass of the Han Dynasty, Journal of Gems & Gemmology, No. 06, pp. 84-92. doi: 10.15964/j.cnki.027jgg.2022.06.008.
2. Liu, S.N. (2022). Study on glass ware of nomads in ancient China, Inner Mongolia Normal University.
3. Huang, X.J. (2023). A Comparative Study of Chinese and Western Ancient Glass in the National Museum of China, Cultural Relics World, No. 03 , pp. 86-91.
4. Hu, Z.Z. & Li, P. & Jiang, L.M. & Wang, T.Y. & Du, G. & Yang, B. (2020). Analysis of LA-ICP-MS components of ancient glass materials and study of their origin, Rock and Mineral Analysis, No.04, pp.505-514. doi: 10.15898/j.cnki.11-2131/ td. 201909210134.
5. Chen, S.Y. & Hou, Z.L. (2019). Analysis of the Ancient Silk Road and Ancient Chinese Glass, China National Exhibition, No. 05, pp. 87-88.
6. Gan, F.X. & Zhao, H.X. & Li,Q.H. & Li,L. & Cheng, H.S. (2010). Scientific and technological analysis and research on the excavated Warring States glassware in Hubei Province, Jianghan Archaeology, No. 02, pp.108-116+151+0.
7. Lu,C. (2007). Influence of the Silk Road on Ancient Chinese Glass Art, Ornamental, No,04, pp. 42-45. doi:10.16272/j.cnki.cn11-1392/j.2007.04.014.
8. Yang, B.D. (1990). A study of the chemical composition of glass formulas of the Qing Dynasty. Journal of the Palace Museum, No. 02, pp. 17-26+38. doi:10.16319/j.cnki.0452-7402.1990.02.004.
9. Shi, M. G. & He, O.L. & Zhou, F.Z. (1986). A Study of Potassium Glasses Unearthed from a Batch of Chinese Han Tombs, Journal of the Chinese Ceramic Society, No.03, pp. 307-313.
10. Zhou, L.Z. (1984). The main factors affecting the weathering of silicate glass. Journal of Dalian Polytechnic University. No. 01, pp.34-44.

Outline Design of an Atmospheric Corrosion Data Monitoring System

Qian Wang*, Zhibin Fan, Baoshuai Du, Lijun Zong, Chunxu Mi, Yaping Wu

State Grid Shandong Electric Power Research Institute, Shandong Smart Grid Technology Innovation Center, Jinan, China

Abstract. Atmospheric corrosion monitoring and evaluation are important means for studying the mechanism and behavior of atmospheric corrosion of metals. In view of the long cycle, cumbersome process, and poor data quality of the traditional outdoor hanging method, an innovative atmospheric corrosion data monitoring system has been designed to detect atmospheric environment data and metal corrosion data on site, and display the detection data of the equipment's service location to users in real-time in the form of a "dynamic corrosion map". This can provide a differentiated design for the anti-corrosion of important equipment such as power grid transmission and transformation engineering components Provide effective data support for engineering life prediction and operation and maintenance. At present, the system has been piloted and promoted in some network provinces, achieving good social and economic benefits.

Keywords: Atmospheric corrosion; corrosion monitoring; corrosion map; data collector

1. Introduction

In recent years, the gradual deterioration of the global climate and the rapid development of industry in China have brought about atmospheric pollution, which has seriously threatened the corrosion of transmission and transformation engineering components and equipment that have been exposed to the atmospheric environment for a long time, affecting the safe operation of power grid equipment. According to statistics, the annual losses caused by corrosion account for about 3% -5% of the global gross domestic product, of which 50% comes from atmospheric corrosion [1-3]. By adopting reasonable anti-corrosion measures, at least 15% of corrosion losses can be avoided [4]. Therefore, monitoring and collecting environmental corrosion data at the service site of engineering components, studying the laws and characteristics of atmospheric corrosion of materials, and rationalizing differential anti-corrosion of power grid equipment are of great significance for improving the operation level of the power grid and reducing economic investment in anti-corrosion. This is also the focus of current research by corrosion workers at home and abroad.

The study on the atmospheric corrosion behavior and regularity of materials is mainly carried out by analyzing the quality changes and morphology of samples during outdoor exposure and indoor accelerated experiments [5,6]. This traditional testing method has drawbacks such as cumbersome operating procedures, long testing cycles, small and discontinuous data, and large errors caused by manual measurement. It also cannot obtain the process information of corrosion over time, which brings certain difficulties to the research of atmospheric corrosion [7]. Atmospheric environmental corrosion detection technology, as an emerging detection technology, has been used to study the corrosion effect of the environment on materials. This method truly reflects the corrosion process of materials in the atmospheric environment by monitoring the strength of corrosion current signals, in order to obtain corrosion data throughout the entire cycle, and achieve continuous automatic monitoring and real-time display of data [8-10]. Atmospheric Corrosion Map is a graphical method for describing the corrosiveness of the atmospheric environment in a target area on a geographical map. Since the 1960s, scholars from various countries have drawn atmospheric corrosion maps of numerous countries, regions, and cities to visually characterize the corrosion characteristics of the regional atmospheric environment [11,12].

The atmospheric corrosion monitoring system innovatively designed in this article is based on the environmental and corrosion data detected by the data collector on site. The supporting intelligent monitoring platform provides corresponding atmospheric corrosion evaluation results in accordance with the ISO 9223-2012 standard and presents them to users in the form of a "corrosion map". This system solves the problems of traditional corrosion map updating cycles being too long and data lagging behind environmental changes, achieving real-time monitoring of atmospheric corrosion and dynamic evaluation of corrosion safety.

* Corresponding author: 190806942@qq.com

2. Overall system architecture design

The intelligent atmospheric corrosion monitoring system consists of three parts: data collection, data processing, and data display. The overall architecture of the system is shown in Figure 1.

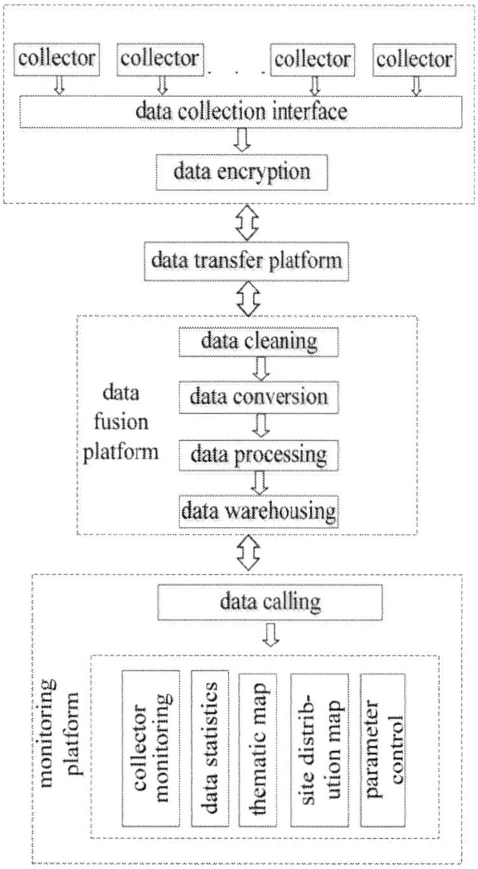

Figure 1. Overall architecture diagram of atmospheric corrosion data monitoring system.

The data collector collects and stores environmental and corrosion data in real time on site. In order to strengthen data security and effectiveness management, it is necessary to encrypt the data and upload it to the data transfer platform, which then transmits the data from the wireless public network to the enterprise information intranet; The data fusion platform in the data processing module preprocesses the data and stores it in the database, including data cleaning, conversion, and simple calculations; The intelligent atmospheric corrosion monitoring platform in the data display module calls the cleaned data, analyzes and processes the data through certain algorithms, and finally presents it to users in the form of thematic maps, chart analysis, and corrosion reports.

The deployment of data collectors should take into account the voltage levels and specific regional needs of substations and transmission lines, follow the principle of equal coverage area, and pay attention to areas with pollution sources.The data collector is equipped with authentication cards, dial-up programs, and channel cards for the operator's VPN (Virtual Private Network) private network that are compatible with the data transfer platform. It collects and stores atmospheric corrosion data in real-time within the region, encrypts the data using national encryption algorithms, and uploads it to the data transfer platform through the VPN private network.

The monitoring platform converts, compares, and analyzes the preprocessed data, and draws a corrosion rate map; According to the ISO 9223-2012 standard, calculate the corrosion levels corresponding to different corrosion rates and draw a corrosion level diagram.

3. Overall Design of Data Collector

The data collector provides source data for the atmospheric corrosion monitoring system, and has extremely high requirements in terms of data acquisition accuracy, transmission safety and reliability, and battery storage capacity. It is mainly composed of five major parts: data acquisition module, data storage module, data control processing module, power management module, and wireless communication module. The overall design diagram of the data collector is shown in Figure 2.

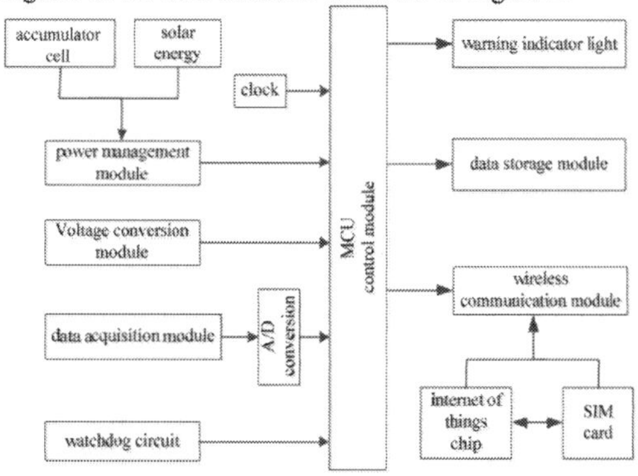

Figure 2. Overall Design Block Diagram of Atmospheric Corrosion Data Collector.

The innovatively designed corrosion sensor can detect the atmospheric corrosion current signal of metal materials when the air humidity is low, and solve the problem of cross contamination of corrosion products between the anode and cathode of the sensor. Compared with traditional corrosion sensors, there are significant improvements and improvements in the sensitivity of data detection and the effective lifespan of sensor use.

The power supply adopts a dual power supply mode of battery and solar panel, with a non solar charging range of more than one month. The motherboard design uses large-scale integrated circuits, multi-layer boards, and surface mount technology, with electromagnetic compatibility meeting strict industry standards and meeting the requirements of outdoor exposure environment work.

The data collectors at each monitoring point collect real-time data on the atmospheric corrosion environment within the region, including environmental temperature, humidity, and corrosion current of metal samples. The

 https://doi.org/10.1051/e3sconf/202447801028

data collection interval can be configured, and the historical data storage time can reach several years. The data collector negotiates keys with the data transfer platform through existing encryption chips, and connects the data to the enterprise intranet. The interactive data is encrypted and decrypted through national commercial encryption algorithms, ensuring safe and reliable data transmission.

4. Design and Effect Display of Atmospheric Corrosion Data Monitoring Platform

The atmospheric corrosion data monitoring platform is based on the State Grid Unified Application Platform (SG-UAP) for software development, achieving communication between the platform and backend services, data transmission, monitoring and management of data collector equipment, and comprehensive display of detection data based on the Geographic Information System (GIS) platform.

The overall design of the atmospheric corrosion data monitoring platform is shown in Figure 3.

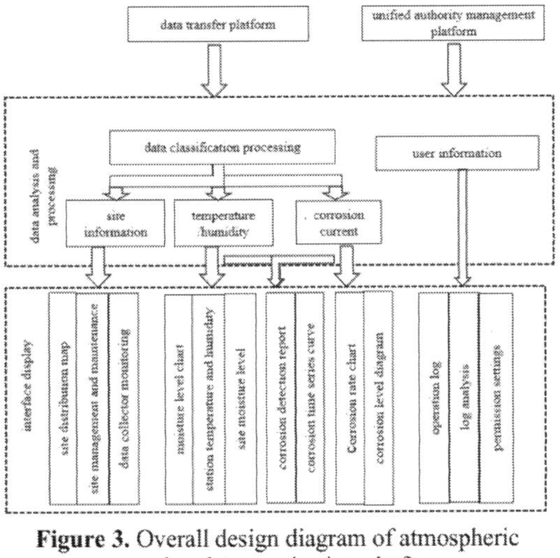

Figure 3. Overall design diagram of atmospheric corrosion data monitoring platform.

The application architecture of the platform is divided into three parts: data management, data analysis, and data display. The data management layer implements hierarchical storage management of data collector ledger information, collected and uploaded raw data, and user login information; The data analysis layer analyzes and calculates the detection data according to standard algorithms, obtaining process data for the data display layer to call; The data display layer calls the process data and presents it in the form of reports, graphics, etc. according to time series and types. The spatial interpolation algorithm is used to fill in the blank area data of the site. According to the principles of ISO 9223-2012 standard, the evaluation results are presented in different colors on the geographic map, and a "corrosion map" with synchronous updates with the detection data is output.

The platform can directly view the geographic location, installation pictures, device status, maintenance records, and full time data of each data collector, and has device anomaly alarm function. Output relevant parameter curves, atmospheric corrosion level reports, and data thematic maps by summarizing, analyzing, and comparing the collected and uploaded detection data. Users can easily and quickly obtain information on atmospheric corrosion in the area of interest by reading the map, in order to determine the severity of corrosion of materials exposed to the atmospheric environment, and provide a basis for formulating corrosion protection strategies for power equipment.

5. Conclusion

The atmospheric corrosion monitoring system developed in this article can comprehensively, meticulously, and accurately grasp the corrosiveness of the natural environment in which the power transmission and transformation project is located. The large-scale application of project results can provide decision data support for corrosion process monitoring, engineering design material selection, and operation and maintenance of in-service equipment. Based on corrosion data, differential design of anti-corrosion measures for engineering equipment can avoid "insufficient protection" caused by inaccurate initial design and excessive investment in subsequent anti-corrosion maintenance. At the same time, it can also avoid waste caused by "excessive protection", greatly improve engineering anti-corrosion efficiency, and reduce energy consumption. The pilot promotion of the system in multiple network provinces across the country has achieved good social and economic benefits.

Material corrosion is a slow and complex process, and there are many atmospheric environmental factors that affect the corrosion process of materials, such as temperature, humidity, SO2, CO, chloride ions, etc. Currently, the environmental data measured by the developed data collector only includes temperature and humidity, and the measured corrosion data is only for metal zinc. The historical accumulation of environmental and corrosion data only represents the corrosion situation that has already occurred in materials or equipment. Discovering or verifying atmospheric corrosion laws from the data, achieving prediction of corrosion data, and future perception of equipment service status can potentially push atmospheric corrosion research to a new level.

Given the above viewpoints, the atmospheric corrosion monitoring system can work in the following two directions in the future: expanding the data detection width on hardware to achieve multi-channel detection of environmental factors such as temperature, humidity, sulfur, salt, and metal materials such as zinc, aluminum, copper, and steel; Conduct data mining research on software in terms of geographic space and time latitude, grasp the corrosion characteristic parameters and main performance changes of commonly used metal materials in power equipment with corrosion development, and achieve prediction of equipment life cycle. Gradually establish a complete corrosion protection technology system and supporting standard system, supporting the development of the selection of anti-corrosion materials towards full life maintenance free direction.

Acknowledgements

State Grid Corporation of China Technology Project (Research on Data Mining and Service Life Evaluation Technology of Power Grid Equipment Based on Atmospheric Corrosion Map of Power Grid 5200-202016471A-0-0-00)

References

1. Belén Chico, Daniel de la Fuente, Iván Díaz, etc.Annual Atmospheric Corrosion of Carbon Steel Worldwide. An Integration of ISOCORRAG, ICP/UNECE and MICAT Databases[J]. Materials, 2017,10(6):601-625.
2. Baorong Hou. Corrosion Cost and Economic Development [J]. China's technology industry, 2020, 02:21-22.
3. Li X, Zhang D, Liu Z, etc. Materials science: Share corrosion data [J]. Nature, 2015, 527(7579): 441-442.
4. Koch G, Varney J, Thompson N, etcl. International measures of prevention, application, and economics of corrosion technologies study [J]. NACE International, 2016, 216.
5. XiaoHe Wang, Mingqiu Wang, Bowen Li, etc. Application of atmospheric corrosion detection technology in the study of atmospheric corrosion of aluminum alloys [J]. Corrosion Science and Protection Technology, 2018,11:671-675.
6. Di Xu, Xiaojia Yang, Qing Li, etc. Progress in Testing Methods and Evaluation Techniques for Atmospheric Environmental Corrosion of Materials [J]. Journal of Chinese Society for Corrosion and Protection, 2022,42(03):447-457.
7. Lin Zhou, Xinpei Yu, Yanan Su. The Application of Sensors in Atmospheric Environment Detection [J]. Chemical Management, 2023(03):60-63.
8. Yanjie Liu, Zhenyao Wang, Binbin Wang, etc. Real time monitoring technology for studying the mechanism of galvanic corrosion under thin liquid film [J]. Journal of Chinese Society for Corrosion and Protection, 2017,37:261-266.
9. Kaiji Liu. Classification and corrosiveness assessment of atmospheric corrosive environments [J]. Overall Corrosion Control, 2015,10:26-27.
10. Xiaoming Wang. Research on the Corrosion Behavior of Zinc in Atmospheric Environment Based on Atmospheric Corrosion Detector [D]. Jinan: Shandong University, 2016.
11. Qihuan Tang, Xianyong Zhang. Typical Atmospheric Corrosion Maps at Home and Abroad and Their Drawing [J]. Equipment Environmental Engineering, 2010,7(4):81-89.
12. Zhibin Fan, Xingeng Li, Xiaoming Wang, etc. Research progress in regional atmospheric corrosion mapping technology [J]. Journal of Chinese Society for Corrosion and Protection, 2023,43(1):29-37.

Research on the Application of Operational Optimization Expert Analysis System Technology

Ziwei Zhong [1,2,*], Lingkai Zhu[1,2], Panfeng Shang[1,2]

[1] State Grid Shandong Eectric Power Research Institute, Jinan, China
[2] Shandong Smart Grid Technology Innovation Center, Jinan, China

Abstract. At present, the national power system is gradually achieving separation of government and enterprise, and corporate restructuring; Implement the separation of network factories and bidding for online access, and gradually establish a modern enterprise system. Effectively carrying out energy conservation and consumption reduction work, reducing the cost of power generation and supply, is an inevitable requirement for the survival and development of power enterprises themselves.To effectively carry out energy-saving and consumption reduction work, as a group power generation company and power plant, it is necessary to solve the following problems: how to evaluate the actual and expected level of unit performance, how to quantitatively determine the impact of various factors on unit performance, how to improve operational level to reduce energy consumption, how to assess and evaluate the implementation of energy-saving work from a management perspective, and so on. Therefore, it is necessary to have a complete and scientific expert analysis system for optimizing power plant operation.

key word: Operational optimization; Expert analysis system; power plant

1. Introduction

Using the basic theory and principles of energy loss analysis, a large number of factors affecting the economic efficiency of the unit were analyzed, compared, and judged. 21 main influencing factors were identified as the basic optimization elements. The relationship between each optimization element was analyzed, and the characteristics of the unit were analyzed. Through dimensional analysis, the overlapping relationship between the optimization elements was briefly analyzed using the superposition method and item by item approximation method.According to the second law of thermodynamics; Improving the initial parameters of the working fluid in a thermal cycle and reducing the final parameters can both improve the thermal efficiency of the cycle. The changes in initial and final parameters during operation will inevitably have a significant impact on the economic performance of the unit. Whether using the heat method or the energy loss method to analyze the thermal economy of a unit, the same conclusion can be drawn: the key to affecting the thermal economy of a turbogenerator unit is the cold source heat loss. The exhaust pressure of the low-pressure cylinder determines the magnitude of the cold source loss. Therefore, the exhaust pressure of the low-pressure cylinder has the most significant impact on thermal efficiency.[1].

2. Thermal economic impact indicators

2.1 Impact of Main Steam and Reheat Steam Temperature on Thermal Economy

The decrease in temperature of main steam and reheat steam, whether it is the absolute ideal efficiency of the cycle or the relative efficiency of the steam turbine, must be reduced.[2]. This is because they reduce the average temperature of the heat absorption process, and at the same time, the decrease in temperature also reduces the specific volume and expansion final dryness of the steam. The former leads to a decrease in absolute ideal efficiency, while the latter leads to steam leakage loss and an increase in wet steam in the flow parts of the high-pressure and low-pressure cylinders of the steam turbine, resulting in a decrease in the relative efficiency of the high-pressure and low-pressure cylinders.

2.2 Impact of Main Steam Pressure on Thermal Economy

When the main steam temperature and exhaust pressure are the same, the change in main steam pressure causes a change in the ideal enthalpy drop of 1 kilogram of steam, which affects the efficiency of the cycle. On the other hand, the change in main steam pressure causes a change in the specific volume of steam and the final dryness of

* Corresponding author: dkyzhongziwei@163.com

expansion, which affects the internal efficiency of the steam turbine.[3]. The specific direction of change should be determined by the specific parameters and range of change.

2.3 Effects of smoke exhaust temperature, oxygen content, air leakage, and environmental temperature

These parameters are the main factors that determine the heat loss of exhaust gas. When the exhaust temperature, oxygen content in the exhaust gas, and air leakage rate (which determine the magnitude of the excess air coefficient) change, the heat loss of exhaust gas and boiler efficiency will correspondingly change.

2.4 Impact of combustible materials in ash and slag

The main parameters that affect the heat loss of incomplete combustion of machinery are fly ash combustibles and slag combustibles, and the latter changes very little during operation, so only the influence of the former can be considered. The increase of combustible materials in fly ash will result in incomplete combustion heat loss of machinery and a decrease in boiler efficiency.

2.5 Impact of Coal Types

It mainly refers to the impact of changes in the calorific value, volatile matter, ash content, and moisture of coal on the economic performance of the unit. In general, the decrease in calorific value and volatile matter makes it difficult for coal to burn completely, increases heat loss during incomplete combustion, and reduces boiler economy. The increase in ash and moisture will make the coal less flammable.

2.6 Impact of water spray desuperheating amount on reheater

The reheater uses water spray to reduce temperature, which completes heat absorption and evaporation at low temperatures, and then completes thermal processes such as expansion and work in the middle and low pressure cylinders of the steam turbine. Compared with the main steam work under subcritical high parameters, the economy is much lower.

2.7 Impact of Boiler Blowdown

The continuous sewage discharge from the boiler is hot water with high temperature, high pressure, and high energy levels. It is gradually absorbed by the supplementary water, heated up, and discharged after reaching saturation temperature in the boiler. It inevitably carries away a lot of heat, resulting in a relatively low thermal economy of the unit.

2.8 Effect of heater terminal difference

The end difference given in the design is determined through economic and technological comparison. Due to various reasons during operation, insufficient heating of the feedwater may increase the end difference, which will increase the irreversibility of heating exchange, generate additional cold source losses, and reduce the thermal economy of the unit. Its impact on economy mainly depends on the size of the end difference and the difference in extraction efficiency between adjacent heaters, as well as the proportion of drainage from the previous heaters.

2.9 Impact of feedwater temperature

The change in feedwater temperature has a significant impact on the economic performance of the unit. It causes changes in the amount of reheated steam extracted, affects the power output, and also changes the heat transfer temperature difference and exhaust gas temperature of the boiler, affecting the efficiency of the boiler. The influence of feedwater temperature on boiler exhaust temperature also leads to changes in boiler efficiency.

2.10 Impact of supplementary water

When supplementary water enters the condenser, without considering the changes in vacuum, regardless of the temperature of the supplementary water, it will not increase work or increase work loss, but it reflects the tightness of the system.

2.11 Impact of condenser subcooling

The subcooling of the condenser increases the loss of cold source and reduces the power output.

2.12 Impact of Efficiency on High and Low Pressure Cylinders of Steam Turbines

The changes in parameters and steam leakage during operation cause changes in the efficiency of the high-pressure and low-pressure cylinders of the steam turbine.

2.13 ower consumption of forced draft and induced draft fans and power consumption of pulverizing system

The power consumption of the forced draft fan and induced draft fan, as well as the power consumption of the pulverizing system, directly reduces the net efficiency of the unit.

Based on theoretical analysis and practical experience, factors have been selected to optimize the economic factors of unit operation. The main parameters that vary independently include: main steam temperature, main steam pressure, reheat temperature, oxygen content in flue gas, carbon content in fly ash, air preheater leakage, environmental temperature, coal type, reheater spray desuperheating, boiler discharge, condensate subcooling, and power consumption of the pulverizing system. Among these parameters, those related to load include: main steam pressure (during sliding pressure operation), air preheater leakage rate, power consumption of forced and induced draft fans, and power consumption of the pulverizing system. The factors affected by other parameters (hereinafter referred to as non independent variation factors) include reheat pressure loss, exhaust pressure, feedwater temperature, make-up water volume, heater end difference, turbine high and intermediate pressure cylinder efficiency, and exhaust gas temperature.[5].

The mutual influence of unit steam parameter deviations includes two meanings: firstly, there is a connection between parameter deviations, such as the main steam temperature deviation causing changes in exhaust flow rate and resulting in exhaust pressure deviation; The second is that the degree to which a certain parameter deviation affects the economic performance of the unit is related to other parameter deviations. Similarly, the heat rate changes caused by the main steam temperature deviation at different exhaust pressures are different. For non independent elements, on the one hand, they can change randomly due to changes in load and environmental temperature, and on the other hand, they are also influenced by multiple factors. Therefore, the superposition method and itemized approximation method are used to solve the influence between parameters (elements).

3. Research on Operational Optimization Theory

3.1 Principle of parameter optimization

3.1.1 Principle of optimizing main steam temperature

The main factor affecting the main steam temperature is boiler combustion adjustment, and the main means of adjusting the main steam temperature include adjusting the boiler combustion and superheater desuperheating water. In theory, the higher the main steam temperature, the more advantageous the economic efficiency of the unit. Therefore, keeping the main steam temperature as high as possible within the safety range specified in the unit regulations is beneficial for the thermal economy of the unit. During the test, the operator first reduces the superheater desuperheating water to the minimum, and then adjusts the boiler combustion. Under the condition of maintaining the main steam temperature stable, the main steam temperature should be as close as possible to the rated temperature. The maximum value that can be reached under this condition is the optimization reference value of the main steam temperature under this condition. From the concept of consumption difference, it is the main steam temperature when the main steam temperature consumption difference+boiler efficiency consumption difference=the minimum value. This experiment needs to be conducted simultaneously with the boiler test due to its involvement in combustion adjustment[6].

3.1.2 Principle of optimizing main steam pressure

The optimization of main steam pressure mainly involves optimizing the opening and throttling loss of the turbine governing valve, as well as the energy consumption of the feedwater pump group. The main influencing factors of main steam pressure are the outlet pressure of the feedwater pump and the opening of the turbine regulating valve. The main measure is to adjust the opening of the regulating valve during constant pressure operation by the operating personnel. When the main steam pressure is close to or equal to the rated value, the efficiency of the high-pressure cylinder of the unit is the highest. When the unit is partially loaded, the main steam pressure is high, which not only increases the throttling loss of the control valve, but also increases the steam consumption of the small unit of the feedwater pump group, and reduces the exhaust temperature of the high-pressure cylinder, thereby affecting the reheat temperature and reducing the thermal efficiency of the unit as in table 1.

Table 1 Determination of optimized main steam pressure values for 135MW units

load	Optimization value of main steam pressure	load	Optimization value of main steam pressure
MW	MPa	MW	MPa
135	13.24	80	7.8
130	13.24	70	7.74
120	13.24	60	7.68

When the unit operates under sliding pressure and the main steam pressure is too low, the thermal cycle efficiency decreases, causing significant economic losses. From the perspective of consumption difference analysis, when the consumption difference of main steam pressure+efficiency difference of high-pressure cylinder+consumption difference of feedwater pump group=minimum, the optimal value of main steam pressure is obtained.

This time, on-site testing was used to make the problem resolution relatively easy.

3.1.3 Principle of reheat temperature optimization

The main factor affecting reheat temperature is boiler combustion adjustment, and the main means of adjusting reheat temperature include adjusting boiler combustion and reheater desuperheating water. Due to the large temperature difference between the desuperheating water of the reheater and the mixture of reheated steam, the irreversible heat transfer loss is significant, and it is equivalent to adding a thermodynamic cycle with medium and low pressure parameters. The impact on the thermal economy of the unit is much greater than that of the desuperheating water of the superheater. Therefore, the control of the desuperheating water of the reheater should be stricter, and theoretically, the smaller the better. Maintaining a high reheat temperature within the safety range specified in the unit regulations is beneficial for the thermal economy of the unit. During the test, the operator first reduces the reheater desuperheating water to the minimum, and then adjusts the boiler combustion. Under the condition of keeping the reheat temperature stable, the reheat temperature should be as close as possible to the rated temperature. The maximum value that can be reached under this condition is the optimal reference value of the reheat temperature under this condition when the reheat temperature difference+the boiler efficiency difference caused by the reheat temperature adjustment=the minimum value.

3.1.4 Principle of reheat pressure optimization

The reheat pressure mainly depends on the pressure loss of the reheat circuit, and the pressure loss depends on the steam flow rate of the unit. For units operating under sliding pressure, when the optimized value of the main steam pressure is determined and the environmental conditions are determined, the reheat pressure and reheat pressure loss are also determined. When the main steam pressure of the unit is adjusted to the optimal value, the corresponding reheat pressure is considered the optimal reheat pressure.

3.1.5 Optimization principle of superheater desuperheating water

The minimum value of the desuperheating water volume of the superheater while maintaining the stability of the main steam temperature and being the optimization reference value is the optimization reference value of the desuperheating water of the superheater. The desuperheating water of the superheater is taken from the outlet of the high-pressure heater, and from the perspective of thermal cycle theory, it has little impact on the economic efficiency of the unit. However, the theoretical application of equivalent heat drop is based on the fact that the expansion process line of the steam turbine remains unchanged. Due to the presence of desuperheating water, the process line will change, and it will also have a certain impact on the economy. At the same time, the temperature difference between the feedwater temperature and the steam temperature of the superheater is very large, resulting in a large heat exchange temperature difference and a large strain force on the equipment. Therefore, efforts should also be made to minimize it.

3.1.6 Optimization principle of reheater desuperheating water

The desuperheating water volume of the reheater, while maintaining a stable reheating temperature and being the optimal reference value, is the optimal reference value for the desuperheating water of the reheater. Due to design reasons, the reheat temperature is relatively low under various operating conditions, and the desuperheating water of the reheater is basically zero throughout the entire test process. From the perspective of thermal cycle, the presence of reheated desuperheating water is equivalent to adding a low-parameter thermal cycle process, which has a significant impact on the economic performance of the unit.

3.1.7 Optimization principle of exhaust pressure

The vacuum of the condenser has a significant impact on the economic performance of the unit, and is the largest of all influencing factors. The determination of the optimal vacuum is influenced by many factors. In theory, the optimal vacuum is the minimum sum of the economic benefits brought about by the increase in vacuum caused by the input of the circulating water pump and the work consumed by the input of the pump, which is the optimal vacuum. At the same time, there are many factors that affect the vacuum of the condenser in the unit. Firstly, it is the heat exchange of the condenser. For the operating unit, the thermal resistance, which refers to the cleanliness of the condenser and the tightness of the negative pressure system, mainly depends on the quality of steam, the quality of circulating water, and the amount of air leakage, and is a gradual process; Next is the environmental conditions, which are the inlet water temperature of the circulating water; Once again, the steam flow rate discharged by the turbine into the condenser corresponds to the load of the unit.

When optimizing the test, the first step is to conduct a tightness test to determine the external working conditions of the condenser, the leakage situation on the negative pressure side, and eliminate all possible leakage points. This is to determine the operating influencing factors of the K value and the operating combination mode of the circulating water pump, as shown in the subsequent circulating pump test. On this basis, conduct tests on condenser back pressure and circulating water temperature, as well as the relationship between condenser back pressure and unit load.

3.1.8 Optimization principle of heater end difference

The main factors affecting the heater include the end difference of the heater, extraction pressure loss, and heat dissipation loss, with the most important being the end difference of the heater. The main factors affecting the end difference of the heater include the size of the heat transfer tube inside the heater, the convective heat transfer coefficient inside the tube, the condensation heat transfer coefficient outside the tube, and the temperature of the medium inside and outside the tube. Calculating the end difference of the heater using heat transfer methods is very cumbersome. For the operating heater, as it has been designed, the factors that affect the end difference are actually the feed water flow rate, the temperature at which the feed water enters the heater, and the steam side temperature at which it enters the heater. Optimize the use of on-site testing methods, adjust the water level of high and low pressure heaters, and check the drain valves of the heaters. For the low-pressure heater, adjust the air extraction pipe and check the temperature rise of the heater. If the terminal difference of the heater is normal, the maximum temperature rise that the heater can achieve is the optimized reference value of the heater.

3.1.9 Optimization principle of exhaust gas temperature

Higher exhaust temperature than the optimized value can lead to a decrease in boiler efficiency. Every 10 °C increase in exhaust temperature can reduce boiler efficiency by 0.5% and increase coal consumption by 1.7g/kWh.

The optimization principle of exhaust gas temperature is to comprehensively consider the influence of various factors, and the minimum temperature that can be reached under the optimal operating state of the boiler is the optimization value. The factors that affect the exhaust temperature include: 1) fouling of the heating surface, coking in the furnace, and ash on the convective heating area at the tail can all cause an increase in exhaust temperature. The boiler should undergo combustion adjustment and timely soot blowing on the heating surface. 2) Boiler combustion. The deviation of the furnace flame, unreasonable coordination of primary and secondary air, high primary air speed, and excessive air volume may cause the flame center to shift upwards and the exhaust temperature to rise. 3) Operation mode of the pulverizing system. Check if the pulverizing system is functioning properly. Focus on checking whether the fineness of coal powder is too coarse, whether the pulverizing air volume is too high, and whether the pulverizing system uses too much cold air. Add steel balls in a timely manner according to the current of the coal mill.

The influence of air supply temperature and air leakage rate of the air preheater. 4) As the supply air temperature increases, the exhaust gas temperature also increases, so the exhaust gas temperature needs to be corrected to 20 degrees below the design ambient temperature.

3.1.10 Optimization principle of oxygen content at the inlet of air preheater

The oxygen content at the inlet of the preheater is one of the important indicators that should be controlled during operation, and it is also the easiest parameter to control and adjust during operation. Other indicators it affects include the carbon content of fly ash and large slag, the total power consumption of the forced draft fan and induced draft fan, the exhaust temperature, and the temperature of the main steam and reheat steam. The analysis and selection of this indicator should be based on the calculation of the power supply coal consumption of the unit before making a comprehensive judgment. The method is to use the power supply coal consumption as the final optimization indicator to determine the optimal oxygen level operating condition that can maintain the economic performance of the entire unit. The ideal oxygen control curve was obtained.

3.1.11 Optimization principle of carbon content in ash and slag

Mechanical incomplete combustion heat loss is one of the main heat losses in boilers, second only to exhaust heat loss. For solid state slag removal coal powder boilers, it is within the range of 0.5-5%. The main factors affecting the heat loss of incomplete combustion of boiler machinery are fly ash combustibles and slag combustibles. In a coal powder furnace, the ash that falls into the cold ash hopper only accounts for a small portion of the total ash amount entering the furnace, and the proportion of ash in the ash in the total ash amount entering the furnace varies very little. Therefore, the mechanical incomplete combustion heat loss caused by combustible substances in the ash is usually only 0.5-1%. The vast majority are caused by combustible materials in fly ash. Therefore, only the impact of the former can be considered. It is generally believed that the residual combustibles in fly ash and ash residue are pure carbon, and mechanical incomplete combustion heat loss can be calculated by measuring the carbon content of fly ash.

The optimal operating condition for a single indicator may not necessarily be the optimal operating condition for the entire unit. Optimization experiments have shown that using low oxygen operation mode can reduce smoke exhaust losses and fan power consumption, and the overall economic efficiency of the unit is the best. However, the carbon content of combustible materials in fly ash and large slag will increase. Therefore, the optimized value of carbon content in ash and slag may not necessarily be the lowest value that can be achieved, but rather the value that maximizes the economic efficiency of the unit.

3.1.12 Optimization principle of total power consumption of forced draft fan

The main factor determining the power consumption of the supply and induced draft fans is the amount of oxygen used during operation. Under the same load, there is a significant difference in total power consumption between the supply and induced draft fans due to the use of different oxygen levels. With the increase of oxygen content, the power consumption of the forced draft fan and induced draft fan significantly increases, leading to an increase in auxiliary power consumption and directly affecting the power supply coal consumption of the unit. The lower total power consumption (rate) of the supply and induced draft fans that can be achieved when operating with lower oxygen levels. Therefore, the optimization value of power consumption for the forced draft fan is determined based on the final optimization index of power supply coal consumption.

3.1.13 Optimization principle of total power consumption in the milling system

The total power consumption of the pulverizing system is mainly determined by: a) the operating characteristics of a single pulverizing system, and b) the operating mode of the pulverizing system under different loads. For the monthly electricity consumption data of the power plant's daily statistics, it is also related to the load curve of the unit this month. When the steel ball loading capacity of the coal mill is constant, the main factor affecting the unit consumption of a single pulverizing system is the output of the pulverizing system. When the pulverizing system operates at high output, the unit consumption of the pulverizing system is significantly reduced. Improving the output of the pulverizing system is an effective measure to reduce the power consumption of the pulverizing system.

3.2 Optimization principle of operation mode

3.2.1 Principle of optimizing the constant sliding pressure operation mode of the main engine

The influence of main steam pressure on the relative internal efficiency of steam turbine units is opposite to that of main steam temperature. When the main steam temperature is constant, as it increases, the specific volume of the steam decreases, and the leakage loss of the steam turbine correspondingly increases and decreases; When the reheating parameters and exhaust pressure remain unchanged, increasing also increases the exhaust humidity, causing an increase in exhaust moisture loss and a decrease. According to the formula, the decrease in relative internal efficiency reduces the effect of increasing. However, due to the increase caused by the increase exceeding the decrease, the overall result is that the actual cycle efficiency of the steam turbine unit is still improved. The existing units of Zhanhua Power Plant are designed with basic load, and 135MW has been operating at constant pressure for a long time. In recent years, with the continuous expansion of power grid capacity and the continuous decrease of unit load rate, it is inevitable for large units to participate in peak shaving of the power grid. In addition, in recent years, there have been renovations to optimize the operation mode of the units, which is to determine the load range of variable voltage operation of the units and the optimal constant and sliding voltage operation parameters. This is extremely urgent. Although increasing the initial parameters can help improve economic efficiency for high parameter and large capacity units, as the main steam pressure increases and the load decreases, the throttling loss of the control valve will increase. Therefore, there will be an optimal value for the main steam pressure under different loads. This optimal value is mainly determined through experimentation. By conducting experiments under different loads and operating modes of constant sliding pressure, the conversion load points of constant sliding pressure are calculated and analyzed.

3.2.2 Principle of power consumption operation mode of circulating water pump

By determining the efficiency of circulating water pumps under different loads; Determine the relationship between circulating water flow rate and power consumption of circulating water pumps under different loads; Determine the relationship between circulating water flow rate and vacuum under different loads; Determine a reasonable operating mode of the circulating water pump to determine the optimal vacuum.

When the inlet temperature of the circulating water is a constant value and the unit load is a constant value, the main factor affecting the pressure of the existing condenser is the circulating water flow rate. In this case, increasing the flow rate of circulating water can reduce the pressure of the condenser, but when the flow rate of the circulating water pump increases, the power of the circulating water pump also increases. Therefore, increasing the flow rate of the circulating water pump is beneficial for the economic efficiency of the unit only when the increased power of the turbine is greater than that of the electric power of the circulating water pump due to the increase in the flow rate of the circulating water. That is to say, when the power generated by the decrease in condenser pressure - the power consumption increased by the increase in circulating water flow=the maximum value, it is the most economical for the unit, and the vacuum at this time is the optimal operating vacuum.

The determination of the optimal vacuum is determined through experiments. The experimental design was conducted separately in different seasons and at different environmental temperatures.

3.2.3 The impact of coal mill operation mode on the overall unit economy

The total power consumption of the pulverizing system is mainly determined by: a) the operating characteristics of a single pulverizing system, and b) the operating mode of the pulverizing system under different loads. The optimization experiment requires comprehensive testing of the operating characteristics of two pulverizing systems. When the steel ball loading capacity of the coal mill is constant, the main factor affecting the unit consumption of a single pulverizing system is the output of the pulverizing system. When the pulverizing system operates at high output, the unit consumption of the pulverizing system is significantly reduced, but for multiple pulverizing systems, the operating characteristics are not the same. Similarly, for the operation mode of the pulverizing system under different loads, increasing the output of the pulverizing system is an effective measure to reduce the power consumption of the pulverizing system.

4. Conclusion

The operation of power plant units is a very complex system, and there are many factors that affect the economic efficiency of the units. The relationship between the influencing factors is very complex. The factors that simultaneously affect the economic performance of the unit can be divided into controllable and uncontrollable factors. And some factors have an impact on both steam turbines and boilers. The ultimate goal of this project is to reduce the coal consumption rate of the unit through experimental adjustments and equipment adjustments by operators, based on the basic principle of "practicality and accessibility". The so-called "practical and achievable" refers to the minimum coal consumption rate or minimum unit heat consumption rate and highest boiler efficiency that can be achieved by the unit after various on-site tests and repeated debugging of relevant equipment by operators. That is to say, the total consumption difference of all parameters on the machine or furnace side that affect the thermal economy of the unit on the unit is the smallest. It is the target value that operators can achieve through their efforts in the operation and mode of the system and equipment, rather than the design value of the unit. The experiment takes the power supply coal consumption rate or unit heat consumption rate of the unit as the final optimization target value, and takes controllable and uncontrollable parameters that affect the unit heat consumption rate and boiler efficiency as optimization variables. When considering the impact of individual parameters on the thermal economy of the unit, other parameters and operating states of the unit are corrected to a comparable benchmark state, in order to accurately determine the impact of parameter changes on the thermal economy of the unit.

Acknowledgements

This work was supported by The Science and Technology Project of State Grid Corporation of China-Research on low-carbon regulation-oriented coal-fired captive power plants' participation in power grid supply and demand regulation and source grid load coordinated operation technology. (5400-202236173A-1-1-ZN)

References

1. K eravnou E T et al, W hat 15 a D eeP E xPert System A n A nalysis of the A x ch iteetu ralR eq u irem ents of Second — G en eratio n ExPert System. The Knowledge Engineering Review, 1989(3): 3-6.
2. Hayes RothE, waterman D. A., Lenat D.B.Establish an expert system. Sichuan Science and Technology Publishing Society, 1986.
3. Zhang Lei. Mathematical Model for Online Monitoring of Thermal Power Units. Master's Thesis of Southeast University, 2006
4. Sun Xintian. Research on the Application of Load Optimization and Distribution System in Thermal Power Plants. North China Electric Power University, 2005
5. Li Xiaojin, Analysis of Consumption Difference of 500MW Units and Its Application in Management of Operational Economic Indicators. North China Electric Power University Xue, Baoding, Hebei. 2003
6. Yuan Guili, Wang Tianhong, Shi Guoqing. Application of Operation Optimization System O PT I PR O in Power Plant Operation Management Research. China Electric Power, 2004, Volume 37, Issue 9.

Reservoir Development Geologic Modeling and Residual Oil Prediction Research

Jinze Fu*

Qingxin Oilfield Co., Ltd. of Daqing Oilfield, Daqing, 151413, China

Abstract. The aim of this thesis is to investigate the geological modeling methods in reservoir development as well as residual oil prediction techniques. Geological modeling is a crucial step in oilfield development, which involves the three-dimensional description of subsurface reservoirs and helps to better understand reservoir properties. Residual oil prediction, on the other hand, is an estimation of the future production capacity of the reservoir and is crucial for decision making and resource management. This study explores different geological modeling methods, including seismic inversion, core analysis, and wellbore data, and introduces a variety of residual oil prediction techniques, including numerical simulation and statistical methods. Through example analyses and case studies, this paper demonstrates the application of these methods in practical reservoir development and their potential to improve reservoir development efficiency and residual oil prediction accuracy.

Keywords: Reservoir characterization; geological modeling; static classification evaluation; reservoir inhomogeneity

1. Introduction

Reservoir development is one of the core aspects of the petroleum industry and involves the extraction of oil and gas resources from underground reservoirs. Geological modeling and residual oil prediction are two crucial aspects in the reservoir development process. Geological modeling aims to create an accurate three-dimensional representation of the subsurface reservoir to help engineers and geologists better understand the nature and structure of the reservoir[1]. Residual oil prediction, on the other hand, is an estimation of the future production capacity of a reservoir, which helps in decision making, resource management and production optimization. This study will explore different geological modeling approaches and residual oil prediction techniques to improve the efficiency and sustainability of reservoir development[2-4].

1.1 Geological modeling methodology

1.1.1 Seismic inversion

Seismic inversion is a commonly used geological modeling method to infer the nature of a reservoir by analyzing the way seismic waves propagate through the subsurface reservoir. This method provides a high-resolution image of the subsurface structure and helps to determine important parameters such as the layers, thickness and porosity of the reservoir[5].

1.1.2 Core analysis

Core analysis is a method of obtaining reservoir properties by laboratory testing of reservoir rock samples[6]. This method provides detailed information about the porosity, permeability, pore structure and fluid properties of the rock. Core analysis is often combined with seismic inversion data to create more accurate geologic models.

1.1.3 Wellbore data

Wellbore data, including wellbore logging and wellbore sampling, are key tools for monitoring wellbore conditions in real time. These data provide information on reservoir properties, the location of the oil-water interface and reservoir pressure[7]. Used in conjunction with seismic inversion and core analysis, wellbore data can help construct more comprehensive geologic models[8].

1.2 Current status of domestic and international research

China's low permeability oil fields are widely distributed, and the four major basins such as Jungar, Songliao and Bohai Bay are rich in low permeability oil reservoirs, which are very rich in resources [9]. As of 2018, according to the statistics of relevant departments, low

* Corresponding author: fujinze@petrochina.com.cn

permeability oil reserves occupy nearly half of PetroChina's proven oil geological cumulative reserves, up to more than 40%, so how to efficiently and economically develop low permeability oil reservoirs has gradually become a worldwide problem[10-12].

1.2.1 Sedimentary feature spreading study

Many scholars in China have also carried out fine studies on deltaic outcrops and modern sedimentation. Li Sitian and other geologists have completed a fine comparative study of the Jurassic Yan'an Formation deltaic outcrops in the Ordos Basin, and at the same time, they have also compared the estuarine dams, mat sands and other sands, which not only established the stratigraphic framework of the basin, but also delineated the internal genesis phases of the sedimentary system. Wang Defa and other geologists carried out research on two deltaic deposits in Daihai Lake, Inner Mongolia, and carried out a comprehensive study based on the theory of "fluvial-phase configuration", and arrived at the technical scheme of dividing the interface of land-phase deltas into 6 levels, and summarized the basic reasons for the composition of the 10 types of land-phase deltaic deposits. Liu Ziliang and other geologists assembled the sedimentary microphase of thick sand body in front of the delta in Longdong area of Ordos Basin, compared the lateral tracking profiles, explored the identification signs of the interface of sedimentary sand body in front of the delta, and classified 9 kinds of constitutive elements[13].

1.2.2 Reservoir fine characterization studies

"Fine reservoir description" technology is based on the theories of sedimentology, tectonic geology, petroleum geology, etc., using computer as a means, comprehensively applying seismic stratigraphy, logging geology, reservoir geology, geostatistics, etc. to qualitatively and quantitatively describe the type of oil and gas reservoirs in three-dimensional space, the external geometric shape, size, reservoir parameters changes and fluid distribution. internal structure, reservoir parameter changes and fluid distribution. The technology of "fine reservoir description" maximally applies computer visualization technology to show the change rule of the reservoir in three-dimensional space, so as to carry out detailed and accurate description of the reservoir, as well as comprehensive and comprehensive evaluation. At the same time, this technology is also an oil field into the development, with the deepening of the degree of exploitation and dynamic data increase, the fine geological characteristics and potential research, as well as constantly improve the process of reservoir geological model[14]. Fine reservoir description is equivalent to the adjustment stage of reservoir management, which is the process of entering the middle and late stages of oilfield development, making full use of modern science and technology and theories, especially computer technology, geology, sedimentology, reservoir geology and other theories on the basis of a large amount of historical data, and serving for the excavation of the maximum potential of the reservoir. Reservoir fine description is a multidisciplinary and comprehensive research on reservoirs, which takes the difficult development unit as the research object, establishes a fine three-dimensional geological model as the basis, reveals the spatial distribution law of residual oil as the focus, and determines the measures for digging out the potential and improving the recovery rate as the ultimate goal, and it is a systematic project integrating multiple disciplines such as geology, logging, numerical simulation, and reservoir engineering, and so on[155].

2. Residual oil prediction techniques

2.1 Numerical simulation

Numerical simulation is a residual oil prediction method based on a mathematical model that predicts future production capacity by simulating fluid flow and rock deformation during reservoir recovery[16]. This method can take into account complex geologic conditions and recovery strategies and provides detailed information on reservoir dynamics. The first step in numerical simulation is to develop a mathematical model that represents the reservoir. This model takes into account the reservoir's geological characteristics, fluid properties, permeability, porosity, and other parameters. Typically, the reservoir is partitioned into a three-dimensional grid and each grid cell contains information about the region. Numerical simulations use the equations of conservation of constant mass, conservation of momentum, and conservation of energy to model the flow of fluids through the reservoir[17]. These equations take into account parameters such as pressure, temperature, and saturation of the fluid, as well as the physical properties of the rocks in the reservoir. In addition to the fluid flow, the numerical simulation also considers the deformation behavior of the rock. This includes factors such as the stress distribution of the rock, fracturing conditions, and porosity variations, all of which affect the release and flow of residual oil[18].

2.2 Modeling and parameterization

In this case, a three-dimensional numerical model has been developed, taking into account geological features, reservoir fluid properties, petrophysical properties and the location of existing extraction wellheads. The model was discretized using the finite element method and computed using conventional simulation software.

2.2.1 Activation function

An activation function is a function that runs on neurons, by which the neural network model is approximated as a nonlinear function, so that the neural network can be applied to numerous nonlinear models. With a linear activation function, the input is passed directly to the output without any correction:

$$f(x) = x$$

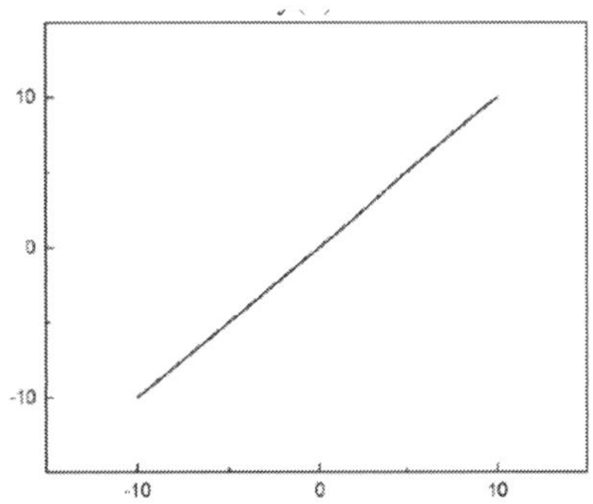

Fig. 1 Schematic diagram of linear activation function

Sigmoid activation function, it will be a sequence of data through the calculation of compression to a large negative to 0 approximation, a very large positive to 1 approximation layer, the value of the output result that corresponds to the predicted probability of the data. Probabilistic prediction model, between 0 and 1, makes it very possible to apply it to the output, the formula is shown below:

$$f(x) = \frac{1}{1+e^{-x}}$$

The disadvantage of the Sigmoid activation function is that it is prone to the phenomenon of gradient vanishing, and the training process converges slowly due to the large training index.

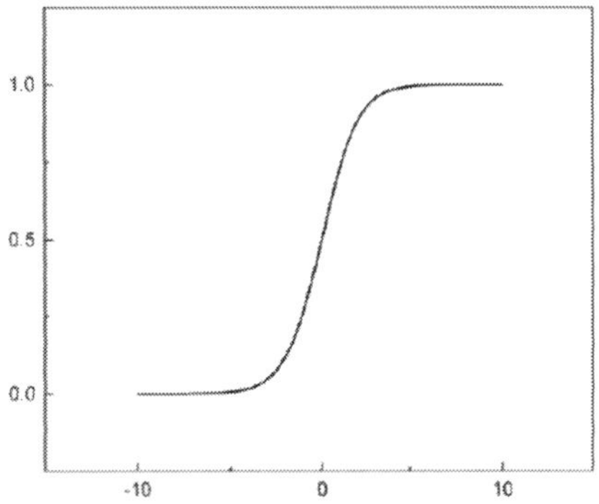

Fig. 2 Schematic diagram of Sigmoid activation function

2.3 Simulation results and residual oil prediction

Through numerical simulation, we modeled the dynamic behavior of the reservoir, including oil production rate, water drive curve, bottomhole pressure and other parameters. Based on the simulation results, we estimated the oil reserves in the remaining reservoir and predicted the future production capacity. The model evaluation indexes are as follows, and the correlation coefficient is used to assess the degree of correlation between the predicted and real values of the model:

$$\frac{\Sigma(y_i - \bar{y})(\hat{y}_i - \bar{\hat{y}})}{\sqrt{\Sigma(y_i - \bar{y})^2}\sqrt{\Sigma(\hat{y}_i - \bar{\hat{y}})^2}}$$

Where n is the number of training variables and the root mean square error of the prediction set RMSEP response model performance on test data:.

$$RMSEP = \sqrt{\frac{1}{m}(\hat{y}_i - y_i)^2}$$

2.4 Analysis of results and decision-making

By analyzing the simulation results, we can optimize the reservoir's recovery plan, including adjusting wellhead production, water injection strategy, and wellhead location. These decisions will help to increase the production capacity of the field and extend the life of the reservoir.

3. Composite function RNN-based oil quantity prediction algorithm for oil wells

3.1 Problem analysis

Oil well production forecasting is a time-series forecasting task, where data from various influencing factors are used to forecast oil production from oil wells of future data for prediction. Among them, RNN has good effect in solving this kind of problems, which is more accurate than traditional statistical learning methods.RNN can better solve the above problems, not only can it take into account the influence of related geologic development factors on the oil production of wells, but also can take into account the correlation of the oil production of wells in the time dimension, which has strong prediction accuracy and strong adaptability to the environment.

3.2 Algorithm Design Principles

The data of the input model in the experiment is represented by a multidimensional matrix, while realized by dot product between the matrix and the matrices. In order to evaluate the accuracy of the oil production prediction model for oil wells, the main purpose is to measure the error between the predicted value and the true value. The smaller the error, the closer the predicted value is to the true value. Root Mean Square Error (RMSE), Mean Absolute Error (MAE) and Mean Absolute Percentage Error (MAPE) are selected as the evaluation indexes, and the calculation formula is as follows[19].

$$RMSE = \sqrt{\frac{1}{n}(y_i - \hat{y}_i)^2}$$

$$MAE = \frac{1}{n} \mid y_i - \hat{y}_i \mid$$

$$MAPE = \frac{100\%}{n} \frac{\mid y_i - \hat{y}_i \mid}{y_i}$$

3.3 Experimental Flow

(1) Processing of oil production data from oil wells, as well as the division of the data set is processed and sliced according to the process in 2.2.

(2) The inputs of both proposed models are pump diameter, pump efficiency, formation thickness, and well oil production.

(3) The training set is trained separately for all models and experiments are conducted using the optimal function of the loss function for all models using the selection of the cost function. And the hyperparameters designed in this paper are continuously optimized empirically so as to perform grid search to get the best experimental results[20].

(4) Save each model and test the unified data, while using the evaluation index in 3.2 to compare and analyze the experimental results, any model in any well are repeated 20 times, in order to ensure the accuracy of the test results, the average value is taken as the final evaluation index results.

3.4 Experimental results of well oil production prediction

Oil well production data is data that shows correlation with time, and its overall trend will show fluctuation with time. Therefore, the multiplicative decomposition model is used. The four data of T1 and T2 wells in T block and H1 and H2 wells in H block are used as experimental data to analyze the basic characteristics of oil production data and the characteristics of each component after STL decomposition. After continuous debugging, it is found that the STL decomposition is most effective at 15, which can ensure that the oil production data of the wells are fully utilized.

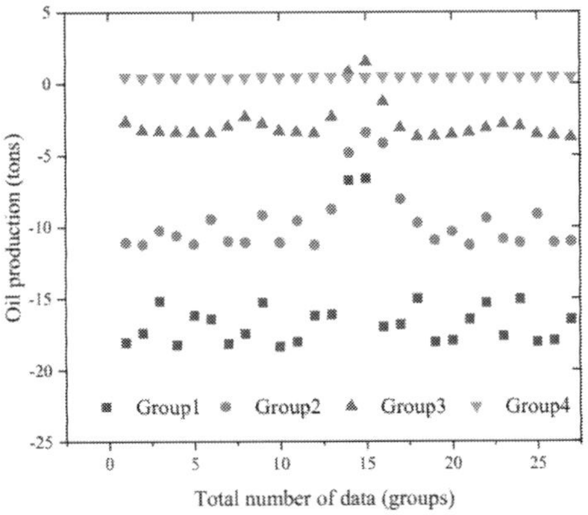

Fig. 3 Schematic diagram of STL decomposition of well T1 in T-break block

Overall, the STL decomposition results in a smoother trend graph, with a more pronounced general trend at a given stage. Relative to the original data, there is much less fine noise and it is changed to a smoother curve. It also has a suppression effect on the singular values at a certain moment. As for the period term, two of the cycles are selected as examples, and it can be seen that the pattern of each cycle is obvious and smooth. Therefore, the trend and cycle terms after STL decomposition reflect the trend and cycle pattern of daily oil production of the four wells more accurately and comprehensively.

4. Conclusion

Geological modeling and residual oil prediction are indispensable aspects of the reservoir development process. This paper discusses different geologic modeling approaches and residual oil prediction techniques, and demonstrates their use in real-world oilfield development through case studies. By building complex mathematical models that take into account geological features and fluid behavior, numerical simulation can provide reliable residual oil estimates and support decision making. The case study in this paper demonstrates the application of numerical simulation in conventional reservoirs, providing a powerful tool and methodology for reservoir development. The STL of time series decomposition is used to decompose the raw well oil production data so that the effect of noise such as residuals can be removed and smoother trend and period terms with stronger correlation can be obtained.

References

1. Application of deep learning in drug activity prediction studies. Liu, Li-Mei; Chen, Xiao-Jin; Sun, Shi-Wei; Wang, Yu; Wang, Hui; Mei, Shu; Wang, Yao-Jun. Advances in Biochemistry and Biophysics,2022
2. Modeling and stability analysis of LCC-S-type wireless energy transmission system based on activation function. Hu Xufang; Wang Yue; Lv Shuangqing; Zhao Delin; Ma Tianlu. Journal of Electrotechnology,2023
3. An attention mechanism-based content popularity prediction algorithm for Bi-GRU. Xu, Xue; Liu, Guangjie. Electronic Measurement Technology, 2022
4. A CNN-GRU based fault diagnosis method for chiller sensor deviation. Li, Donghui; Zhao, Mozhuan; Gao, Long. Journal of Hunan University (Natural Science Edition), 2022
5. A review of deep learning-based fault diagnosis and prediction methods. Peng, Cheng; Li, Fengjuan; Jiang, Jinyuan. Modern Electronic Technology, 2022
6. An unbalanced data classification method based on cost-sensitive activation function XGBoost. Li, Jingtai; Wang, Xiaodan. Computer Science,2022
7. Deep learning based on RNN-LSTM neural network for driverless cart usage environment and data processing. Fan, J.Y.; Xu, C.L.; Gao, CH. Electronic Fabrication,2021
8. Oil production prediction of oil wells based on long and short-term memory networks and attention mechanism. Pan, Shaowei; Zheng, Zechen; Wang, Jizhe; Cai, Wenbin; Wang, Chaoyang. Science, Technology and Engineering,2021

9. A background classification method for offshore target detection using CNN. Yannan Xu; Ningbo Liu; Hao Ding; Key; Yong Huang. Journal of Electronics,2019
10. Application of GA-LSTM model in highway traffic flow prediction. Wen Hui-Ying; Zhang Dong-Ran; Lu Si-Yuan. Journal of Harbin Institute of Technology,2019
11. Comparison of methods based on machine learning for oilfield production prediction. Lin, Xia; Wu, Boyu; Wang, Hongliang; Liu, Zongshang. Information System Engineering,2019
12. Research progress of attention mechanism in deep learning. Zhu, Zhangli; Rao, Yuan; Wu, Yuan; Qi, Jiangnan; Zhang, Yu. Journal of Chinese Information, 2019
13. Analyzing the current situation of petroleum extraction and its measures in China. Wu Hao. Chemical Management,2019
14. Simulation and forecasting of rainfall runoff based on long and short-term memory neural network (LSTM). Yin, Zhaokai; Liao, Weihong; Wang, Ruojia; Lei, Xiaohui. South-North Water Diversion and Water Resources Science and Technology,2019
15. A review of machine learning-based methods for equipment remaining life prediction. Pei, H.; Hu, C. H.; Si, S. S.; Zhang, J. X.; Pang, Z. N.; Zhang, P.. Journal of Mechanical Engineering,2019
16. Arrhythmia recognition of ECG signals based on deep residual convolutional neural network. Li Duan; Zhang Hongxin; Liu Zhiqing; Huang Juxiang; Wang Tian. Journal of Biomedical Engineering,2019
17. A data mining-based method for oil well production prediction with long and short-term memory network model. Gu Jianwei; Zhou Mei; Li Zhitao; Jia Xiangjun; Liang Ying. Specialty Oil and Gas Reservoir, 2019
18. Short-term traffic flow prediction based on KNN-LSTM. Luo Xianglong; Li Danyang; Yang Yu; Zhang Shengrui. Journal of Beijing University of Technology,2018
19. Well production prediction based on ARIMA-Kalman filter data mining model. Gu Jianwei; Sui Gulei; Li Zhitao; Liu Wei; Wang Yike; Zhang Yigen; Cui Wenfu. Journal of Shenzhen University (Science and Technology Edition),2018
20. Research on deep learning optimization algorithm. Tong Weiguo;Li Minxia;Zhang Yike. Computer Science, 2018

AUTHOR INDEX